OUR COSMIC FUTURE

Shall we return to the Moon? Could we colonise Mars, and other planets in our Solar System? How might we travel to the distant stars, in our own Galaxy and beyond? Why haven't we yet met an extraterrestrial civilisation? How can we avoid the various cosmic threats, such as asteroid collisions? Could we escape the remote but certain death of our Sun? What is the ultimate fate of the Universe itself?

This captivating and unprecedented book is about the future of the human race in the Universe for the centuries, millennia and eons to come. It is not an account of 'what will happen', but of 'what could happen' in the light of our current knowledge and scientists' speculations, and its philosophical and social implications. Drawing also on historical accounts and classic works of science fiction, this book artfully displays a gripping preview of *Our Cosmic Future*.

NIKOS PRANTZOS is a researcher at the Paris Institute of Astrophysics, specialising in stellar nucleosynthesis, galactic evolution and high-energy astrophysics. He has written dozens of academic papers, as well as popular-level articles and two other popular science books. The French edition of this book, entitled *Voyages dans le futur*, won the 1998 Jean Rostand prize given by the association Mouvement Universel de Responsabilité Scientifique.

OUR COSMIC
FUTURE

Humanity's fate in
the Universe

NIKOS PRANTZOS

Translated by Stephen Lyle

CAMBRIDGE
UNIVERSITY PRESS

149419

PUBLISHED BY THE PRESS SYNDICATE OF THE UNIVERSITY OF CAMBRIDGE
The Pitt Building, Trumpington Street, Cambridge, United Kingdom

CAMBRIDGE UNIVERSITY PRESS
The Edinburgh Building, Cambridge CB2 2RU, UK   http://www.cup.cam.ac.uk
40 West 20th Street, New York, NY 10011-4211, USA   http://www.cup.org
10 Stamford Road, Oakleigh, Melbourne 3166, Australia
Ruiz de Alarcón 13, 28014 Madrid, Spain

First published in French as *Voyages dans le futur*
English edition published 2000

Printed in the United Kingdom at the University Press, Cambridge

Typeface Monotype Sabon 11/13 pt.   *System* QuarkXPress™ [SE]

A catalogue record for this book is available from the British Library

Library of Congress Cataloguing in Publication data

Prantzos, Nikos.
[Voyages dans le futur. English]
Our cosmic future : humanity's fate in the universe / by Nikos Prantzos ; translated by
Stephen Lyle.
p.   cm.
Includes bibliographical references.
ISBN 0 521 77098 X (hardback)
1. Outer space – Exploration.   2. Interplanetary voyages.   I. Title.
QB500.262.P7313   2000
919.9′04–dc21   99–053248

ISBN 0 521 77098 x hardback

*To Konstantin Tsiolkovsky, Herbert G. Wells, John D. Bernal,
Olaf Stapledon, Arthur C. Clarke, Freeman Dyson and all explorers of our
cosmic future: hoping that it will be much longer, more complex and richer in
events than anything they have been able to imagine.*

CONTENTS

FOREWORD

Our part of the Universe has been an area of active exploration for four decades now. The Moon has received our visit and instruments have been set up on Venus and Mars. Probes have ranged over the Solar System from Mercury to beyond Pluto. Many further projects are in preparation, notably investigation of Saturn's largest moon, the mysterious Titan, its atmosphere brimming over with carbon and nitrogen, the basic ingredients for terrestrial life.

Sometimes, long before any laboratory studies and launch pad engineering, science fiction writers have been at work in the same field. Storytellers like Jules Verne and H.G. Wells have constructed scenarios which bear an astonishing resemblance to later realities. Their intuitions have acted as catalyst in many projects and their fertile imagination has often been a starting point for the careers of astronauts and engineers.

We may well wonder what the future holds in store. How far will we be able to go in our cosmic investigations? As far as the stars? To distant galaxies? Will the strange bestiary of pulsars, quasars and black holes be within our reach one day? What projects are already feasible with today's instruments? And with tomorrow's technology? What are the limitations laid down by theoretical considerations concerning the behaviour of matter? Are such limitations insurmountable?

Internationally renowned astrophysicist Nikos Prantzos is a first class guide in this field. A careful and enthusiastic science fiction reader, he explores various scenarios for the future, bringing his professional skills to bear. He assesses their relevance in the short and longer terms, describing the state of the Universe over the billions of billions of years to come.

We must remember that these futuristic scenarios are entirely based upon the present state of our knowledge. But as further research is carried out, so science advances. And new knowledge can always change the predicted course of future events.

There is no shortage of historical examples. At the beginning of the nineteenth century, remarkable progress in celestial mechanics had given force to a belief in the absolute determinism of natural phenomena. The future was, at least in principle, entirely predictable. Nothing 'new' could happen. 'Freedom and inventiveness were stubborn illusions', as Einstein was later to remark. The future was irrevocably condemned to endless and monotonous repetition.

In the twentieth century, two new ideas came to challenge this desolate conception of the future: quantum physics and so-called deterministic chaos theory. Under their guidance, chance and unpredictability were reinstated. Original and creative thought was once more possible, to the relief of all.

Another example has arisen from developments in thermodynamics. A wide range of temperatures can be observed in the Universe today. Hundreds of millions of degrees separate the burning cores of stars from the chill of cometary ices. Concluding from the irresistible increase of entropy, physicists at the close of the nineteenth century were predicting that temperatures must one day attain total uniformity over the whole Universe. The consequences for life were unavoidable: it could only disappear.

However, studies of the way gravity could affect entropy, together with the discovery that the Universe is expanding, have forced futurologists to revise their copy books. Far from being whittled down, the temperature differences mentioned above tend to increase as time goes by. It is essential to bear in mind this close relationship between the necessarily temporary state of scientific knowledge and the pictures we make of our future.

Nobody is in a position to answer the key question: where do life and consciousness stand? Are they condemned to disappear, or is there some way they can perpetuate their existence indefinitely?

The discovery of nuclear forces less than a century ago revealed a vast store of energy previously unknown to us. What other forms of energy remain to be uncovered, which could prolong our stay in this world?

The extraordinary advances of contemporary science should not lead us to forget that the future remains largely open.

Hubert Reeves

ACKNOWLEDGEMENTS

Many colleagues and friends have helped me with their comments and encouragement throughout the process of writing this book. I would particularly like to thank Sylvie Vauclair at the Laboratoire d'Astrophysique de Toulouse, Sylvie Cabrit at the Paris Observatory, Anne Lefebvre at the Centre de Spectrométrie nucléaire in Orsay, Evry Schatzman and Ludwik Celnikier at the Meudon Observatory, and Stéphane Arnouts, Jean Mouette, Daniel Kunth and Parviz Merat at the Institut d'Astrophysique in Paris. Special thanks should go to Hubert Reeves for his advice and also to Jean-Marc Lévy-Leblond for his critical and patient reading. And finally, I am grateful to Guy Paulus, a long-standing friend at the Université Libre de Bruxelles. Without his invaluable help, I could never have put together this book.

INTRODUCTION

The American anthropologist Ben Finney defined 'man' as an 'exploring animal'. Today, exploration of our own world, Earth, has largely been completed. There is almost no corner of the planet in which humans have not ventured. Only the ocean bed has kept its secrets, and this will no doubt continue for some time to come.

The next step for mankind may well be space, the 'final frontier', as announced in the opening passage of the famous television series Star Trek, which is still stirring imaginations the world over. Human fascination for space, the desire to raise themselves amongst the stars, has been a constant theme since the dawn of humanity. The legends of Icarus and the Tower of Babel are witness to this obsession. However, both these undertakings ended in tragic failure. Clearly, God (or Nature?) would not tolerate excessive human arrogance and ambition.

It was only at the turn of the twentieth century that mankind finally stumbled upon the keys that could unlock the doors to space. The Russian Konstantin Tsiolkovsky realised that the only way to move through empty space is by rockets, based on Newton's principle of action and reaction. But this herald of space travel went much further. Lodged midway between science and science fiction, his ideas were ambitious indeed: permanent residence in space, colonisation of other planets in the Solar System, and even trips to distant stars in order somehow to procure their energy, whenever our own Sun should fade. According to him, stars are the destiny of our species: 'Our planet is the cradle of intelligence', he wrote, 'but one cannot live forever in a cradle'.

Tsiolkovsky's prophesy began to take shape half a century later, in a context he would no doubt never have imagined. The space race had become one of the high stakes of the cold war between two superpowers that emerged after the Second World War. The USSR was first to send a satellite (Sputnik), and then a man (Yury Gagarin), into orbit around the Earth. But the Americans were first to take a person to another heavenly body. On 21 July 1969, Neil Armstrong made 'one small step for man, one giant leap for mankind' when he set foot on the Moon.

This conquest of our natural satellite turned out to be a rather expensive enterprise, and without real interest at the time, apart from the national prestige it entailed. Despite American and Soviet determination to send men to Mars before the end of the century, no human being has ventured further than a few hundred kilometres from the cradle since 1973. This was the date of the last American mission to the Moon, and also the advent of the oil crisis, which marked the beginning of a long period of stagnation in the world economy. It was clearly no coincidence. At the beginning of the twenty-first century, humanity is confronted with serious problems, and pessimism is the order of the day. Economic slump, the population explosion, depletion of resources and pollution leave little room for cosmic dreams.

Paradoxically, some would consider human inability to solve problems on Earth as a motivation for fleeing the cradle, to colonise space and finally to set up the ideal community elsewhere. This utopian thinking, whereby people successfully overcome problems in interstellar space that they have been unable to deal with on the reduced scale of their own home, Earth, clearly manifests a certain incoherence. For others, space travel is not motivated by escape, but rather by the basic urge of the 'exploring animal' to range endlessly over new territories, in search of new resources and new knowledge. Yet others, sharing Tsiolkovsky's prophetic vision, would simply say that we have no choice, that the stars are indeed our destiny. The *homo spatialis* stage in human development may be just as important as the *homo faber* stage in the long process of hominisation. In the words of Edward Young, in his poem *Night Thoughts*, 'Too low they build, who build beneath the stars'.

Will we travel to the stars one day? And if so, how and to what end? What will be the fate of mankind in space in the coming decades, centuries and millennia? Might we meet some other form of life, possibly a kindred spirit, elsewhere in the Universe? Or are we condemned to cosmic solitude? In the much longer term, what does the future hold for Earth, the Sun, our Galaxy, and even the whole Universe? What is the human place in the evolving Universe revealed by modern cosmology? Will the Universe come to an end, as predicted not only by millenarian eschatology, but also by nineteenth-century science? Or will life and intelligence continue for ever? These are the questions addressed in this book in the light of today's knowledge and understanding. It aims to investigate our cosmic future in the medium term (on a scale of centuries), the long term and the infinitely long term, where the latter refers rather to present limits in our ability to extrapolate into the future.

The dangers involved in any attempt at futurology are well known. This is clearly illustrated by a sketch of Paris made in the nineteenth century. The Eiffel tower is shown in the year 1940, surrounded by myriad flying machines, whilst not a single car is to be seen in the streets below! Of course, social and economic factors, so very unpredictable even in the short term, are far more significant than technical or scientific aspects in this kind of futuristic extrapolation. But should we therefore forgo any attempt at long-term conception of the future? This is not my own view. Tsiolkovsky's vision, and those of so many others before him, have shown that this ability to dream about the future is vital for the human race. It is a way of opening up new directions. Although it has no predictive power, it is a prospecting skill which can nevertheless shape the future in our collective imagination, even if only partially and indirectly, and even if that future remains by definition unpredictable. Moreover, the will to conceive of a future beyond nearby temporal horizons is a sign of youth. It is the prerogative of children and teenagers to dream of their future, ever conscious that an accident may bring their dreams to a premature end. But they are unable to prevent themselves. It is only much later that dreams of a distant future begin to fade, and with reason. For should we judge that humanity has already reached the grand old age when its days (or centuries) are numbered? Without wishing to appear excessively optimistic, I cannot say that I hold this view.

This book is not an essay about what is going to happen. It aims rather to suggest what might happen, on the basis of current knowledge and projects, or just in terms of current ideas among scientists. Apart from the feasibility or potential usefulness of these projects, it is interesting to see in what way contemporary science can give substance to the age-old dream of visiting the stars, what perspectives science opens up for the utopian ideal and what kind of answers it may bring to eschatological questions.

The first chapter deals with some current projects concerning colonisation of our immediate neighbourhood in space, e.g., the Moon, Mars, the asteroids, and then the whole of the Solar System. Certain of these projects already raise questions of cosmic ethics, which our species must confront sooner or later. The second chapter is devoted to the next step in our conquest of space, namely interstellar travel (fast or slow). This is an undertaking which is likely to prove extraordinarily difficult. Moreover, the idea that we may one day engage upon interstellar travel raises a particularly interesting question. Any civilisation which had already acquired this ability would spread across the Galaxy in a relatively short time on cosmic scales. The fact that there are no traces of extraterrestrial life in our Solar System might then imply that we represent the most advanced technological civilisation in our Galaxy. The third chapter discusses the very-long-term future of humanity in the Solar System. It is quite likely that our descendants will be faced with cosmic disasters which endanger the very survival of our species on Earth. The most serious will be the death of the Sun itself. And finally, the fourth chapter deals with the very-long-term future of the Universe. Modern cosmology has revealed a Universe in evolution, in which it will be hard for intelligence to survive for all eternity.

Throughout the text, there will be many references to the literature of science fiction, 'the only true literature today', in the view of Jorge Luis Borges. Without completely sharing his enthusiasm, I believe that such scientific anticipation has today become an accepted literary form, although long despised in literary circles. The kind of problems and the way they are treated in this literature have often been a source of inspiration to scientists. There is certainly a strong connection with many of the subjects discussed in this book, as the reader will no doubt observe.

NEAR FUTURES

Our planet is the cradle of intelligence, but one cannot live forever in a cradle.

Konstantin Tsiolkovsky

It is difficult to say what is impossible, for yesterday's dream is today's hope and tomorrow's reality.

Robert P. Goddard

Three hundred thousand trillion kilometres from the centre of our Galaxy lies an inconspicuous yellow star. It appeared relatively late in the majestic disk of the Milky Way, in an age when many first-generation stars had already died. Since then, it has tirelessly followed its circular path around the galactic centre, in the same way as a hundred billion other stars. Despite a quite prodigious speed of roughly 800 000 km/hr, it takes 225 million years to cover its immense orbit. Indeed, since its birth 4.5 billion years ago, it has only completed the round trip about twenty times. Around it is a genuine mini universe, our Solar System, a cortege of about ten planets and myriad smaller bodies, which follow faithfully through this long journey.

Among all the bodies in the Sun's family, only the third has been witness to a form of life evolving on its surface. Over millions and billions of years of evolution, more and more complex beings spread through the oceans, then the land and then the air of this planet. A few million years ago, some of these beings gradually began to adopt an upright position, using their front limbs to handle tools and weapons, and developing a vocal form of communication amongst them. These novel skills allowed them to dominate all other species, and occupy almost all unsubmerged regions of the planet.

Raising their eyes to the night sky, these beings observed countless points of light moving around the centre of their Universe, which was Earth. They were long intrigued by these lights. Were they divine beings in their celestial kingdom? Or were they just ordinary pieces of molten rock? Could they be holes in some dark veil, allowing only a glimpse of the divine fire surrounding the Universe? And if, perchance, there were lands similar to their own somewhere between those points of light, could they too harbour beings able to admire this same sky? Despite their fear of the unknown, they would certainly have liked to go nearer and observe those shining points from close at hand.

Little by little, the bipeds abandoned their myths. They realised that neither their Earth nor even their Sun was the central point of the Universe. They also understood that the points of light were in fact distant suns, lying at much greater distances than their ancestors could ever have imagined. As Earth exploration went ahead, uncharted territories receded, leaving less and less room for dreams of adventure. The vastness of interstellar regions thus became all the more attractive to those who dreamed of new worlds, utopians who sought to build their ideal societies as far as possible from the tyranny, corruption and misery so rife on their own planet.

Only recently have human beings begun to achieve the ancient dream of space travel. Just fifteen years after the greatest massacre in the history of the species, they finally succeeded in freeing themselves from the grasp of Earth's gravitational attraction. Then for the first time they could view the cradle of their species from the outside. A few years after cutting the umbilical cord with mother Earth, they were able to walk upon the nearest of the heavenly bodies and send unmanned probes to investigate the distant confines of the Solar System. Today, at the dawn of the third millennium, they are already planning the next stage of their cosmic adventure.

From dream to reality

The idea of space travel appeared early on in the history of humanity. The first writers on the subject were less concerned with the means of transport than with allowing a free course to their imagination. Thus, in 167 AD, Lucian of Samosata sends the hero of his *Icaromennipus* to the Moon with the wings of a bird, whilst in *True History* (actually a collection of lies), the hero's boat is lifted into the skies by a great storm. Much later, at the beginning of the Renaissance, the Italian poet Ariosto sends Astolfo to the Moon on the back of a hippogriff to seek the lost reason of the hero of *Orlando Furioso*. Means of locomotion presented no greater complications to Johannes Kepler. Although a recognised scientist, he did not hesitate to call upon the services of a genie in *Somnium (The Dream)*, when dispatching his hero to the Moon (the favoured destination of space travellers for a long period). It was only in 1655 that more sophisticated means were envisaged by Cyrano de Bergerac. In *The States and Empires of the Moon and Sun*, his hero devises a system using rockets, lit one after the other, to wrench him from the terrestrial attraction. This did not prevent him from employing methods that were distinctly more farfetched on other occasions.

Adventure and exploration were not the main motivation of these writers when inventing their imaginary travels. Often enough, they are a mere pretext for social criticism, or for presenting a certain vision of the world. In *Micromegas*, Voltaire plays guide to an inhabitant of Sirius. From his great height of 120 000 feet, the visitor observes the absurd behaviour of human beings with a truly astonished eye. In this work, Voltaire sought to attack the scholastic stance of Saint Thomas Aquinas. The latter, influenced by Aristotle, had placed mankind at the centre of the world in his *Summa Theologica*. Voltaire may have been inspired by *Gulliver's Travels*, in which Jonathan Swift delivers a forceful criticism of the society of his day.

Space adventures were not an essential ingredient for those who wished to criticise society and put forward plans for a better world. Until the end of the eighteenth century, sufficient unknown and distant lands remained across the globe to shelter the towns and countries of the utopian dream. Following the way indicated by Plato in *The Republic*, writers like Tommaso Campanella in *The City of the Sun*, Thomas More in *Utopia*, and dozens of others with them, put forward their vision of an ideal society, freed from injustice and poverty by reason or technical advance.

Among the utopians, Francis Bacon was undoubtedly the author whose writings had the biggest impact on Western civilisation. In his masterpiece *New Atlantis*, published in 1627, he imagines a society guided by science, technology and trade, organised in such a way that all can benefit from progress. The island of Bensalem, discovered by the hero after crossing the ocean, bears much closer resemblance to today's Japan than to Elizabethan England. Its citizens have at their disposal refrigerators, airplanes and submarines. 'Merchants of Light', who are a kind of modern industrial spy, travel through distant countries in search of technical information which might be of benefit to Bensalem. They are sent and guided by 'Interpreters of Nature', pure scientists and true rulers of the country, who fix the main lines of research and development.

The idea that science may bring long-term technological and material benefits seems obvious today, provided research is well-organised and freed from any religious interference. But in the seventeenth century it was truly revolutionary. Solomon's House, the technical and scientific hub of Bensalem, prefigures today's academies of science and research centres. The immense impact of Bacon's work is explicitly recognised by Diderot and d'Alembert, who pay homage to him in their monumental work *Encyclopédie*, which appeared in 1751. According to Bertrand Russell, it is to Bacon that we may attribute the maxim 'Knowledge is power'.

The explosion of science and technology throughout the eighteenth and nineteenth centuries significantly altered our view of space. At last, ways of reaching space aimed to accord with the technological knowledge of the day. Edgar Allen Poe, the enfant terrible of nineteenth-century American literature, used a hot-air balloon to transport Hans Pfaal to the Moon. Far better known is the extraordinary journey of Jules Verne's heroes in *From the Earth to the Moon*. At the time of the industrial revolution, Verne imagined using a cannon 300 m long, 'The Columbiad', to blast his space vehicle into orbit around our natural satellite. In fact, he was perfectly aware that this was unrealistic, despite the apparent sophistication of his machine. The passengers would have been killed outright by the terrible acceleration or the heat due to friction with the atmosphere after leaving the cannon.

The other giant figure in science fiction during the late nineteenth century was Herbert G. Wells. He envisaged a rather more esoteric means for sending *The First Men in the Moon*. In this short story, published in 1901, effects of gravity are cancelled by use of 'cavorite', a material invented by Professor Cavor. Inventor and assistant thereby accomplish the first space trip to be brought to the silver screen. Indeed, inspired by Wells' work, the Frenchman Georges Méliès produced a fifteen-minute film in 1902, entitled *A Trip to the Moon*.

Despite their attempts at realism, these heralds of the science fiction age (Poe, Verne, Wells) were suggesting means of transport as laughable as those put forward by Lucian or Ariosto. In order for the dream to become reality, another way had to be found. The first to discover the key which would open the door to space travel was Konstantin Tsiolkovsky. This Russian schoolteacher, working alone in the little village of Kaluga near Moscow at the turn of the century, realised that the only way to obtain motion in empty space was the rocket. Tsiolkovsky established his celebrated equation for rocket propulsion in 1897 and presented it in 1903, in his work *Exploration of the Universe with Reaction Machines*. He even had the idea of replacing gunpowder, the only fuel considered feasible at the time, by a mixture of liquid hydrogen and oxygen. This would allow a considerable gain in efficiency. He also conceived of a multistage rocket and imagined space suits to protect from the cold and the vacuum in space. However, his visions went much further than this. He envisaged agricultural development in space colonies orbiting Earth, and the use of solar energy for space travel. In his book *Dreams of the Earth and Sky*, published in 1895, he noted that Earth intercepts only one billionth of the energy radiated by the Sun. He suggested that the human race could take advantage of the whole of this inexhaustible energy supply, if it could only succeed in colonising the rest of the Solar System, starting with the asteroids because these are small bodies and easy to control!

Tsiolkovsky's Promethean visions were in accord with the idea of the 'new man' promoted by the Bolshevik revolution. As a result, he was elected a member of the Soviet Academy of Sciences in 1918 and received many honours before his death in 1935. He is recognised worldwide for his precursory role in astronautics, stated so clearly by his German counterpart Hermann Oberth in their correspondence: 'You lit the flame. We shall not let it die. We shall set out to achieve man's greatest dream.'

However, rather than serving human dreams, the emergent science of astronautics almost brought about human destruction. In 1943, at the secret base of Peenemünde in northern Germany, Wernher von

Braun and his team began to develop the famous V2 rockets (an abbreviation for Vergeltungswaffe 2, or Vengeance weapon 2). This was Germany's final attempt to change the course of the war. Several thousand V2 bombs fell on England, causing 2500 civilian victims. But it was already too late. The fate of the Third Reich had been sealed. Many German scientists and technicians were subsequently forced to take part in Soviet and American space programmes. The main aim of these programmes was to construct a fleet of intercontinental missiles able to despatch nuclear warheads into enemy territory. Fortunately, the cold war engendered less formidable forms of competition. Amongst these the space race is undoubtedly the most prestigious.

Glory came first to the Soviet Union. On 4 October 1957, the centenary of Tsiolkovsky's birth, the first artificial satellite (*sputnik*, in Russian) was put into orbit around Earth by the later legendary Sergei Korolev and his team. This was soon followed by an impressive series of successes. Sending Yury Gagarin into Earth orbit was undoubtedly the Soviets most symbolic achievement. On 12 April 1961, he became the first man to leave the cradle of our species. The Vostok capsule made one trip around the planet in 108 minutes, at an altitude of about 100 kilometres.

The Americans were taken aback at first by this show of Soviet technological supremacy. However, it was not long before they were to react. In a famous speech in 1961, President John F. Kennedy laid out his country's ambitions to send the first man to the Moon by the end of the decade. The Apollo programme was thus born. On 21 July 1969, Neil Armstrong entered the annals of history, making the first ever human footprints on another heavenly body. All those earlier dreams, of Lucian, Kepler and Cyrano, had at last been realised.

The Americans returned to the Moon seven times before their last mission in 1973, when the exorbitantly expensive Apollo programme finally came to an end. Following this they set up the space shuttle programme, which began in the 1980s and brought many successes, but also the dreadful tragedy of Challenger. Meanwhile, the Soviets had opted for development of space stations (Salyut and Mir). Their main aim was to study the behaviour of the human body over long stays in space.

Despite the present economic slump, it is likely that humans will return to the Moon during the first or second decades of the next century. Indeed, the Moon will be an essential step in their conquest of the Solar System.

Magnificent desolation

The Moon is Earth's closest neighbour, 384 000 kilometres away. It was indeed this proximity which made it a favoured destination for the space travellers of Lucian, Ariosto, Cyrano and other dreamers of the past. The very idea of space travel would surely not have arisen so early in the history of mankind, had it not been for our nocturnal companion.

Between 1959, the date of the first flyby by the Soviet probe Luna 1, and 1976, which marked the return of Luna 24, a total of 56 missions were devoted to the exploration of our natural satellite. This undertaking culminated in the lunar landing of six American teams between 1969 and 1973, which brought 380 kg of lunar material back to Earth. Following this first wave of explorations, only three further probes, one Japanese and two American, have gone into orbit around the Moon over the past twenty years.

The Moon has a radius slightly greater than one quarter of Earth's radius and an area of 40 million square kilometres, comparable with the area of the American continent. It has mass 83 times smaller than Earth, and gravity at its surface is six times weaker than it is at the surface of our own planet. The lunar atmosphere is more tenuous than the best vacuum obtainable on Earth. It has a density hundreds of trillions of times lower than Earth's atmosphere and its total mass does not exceed about 10 tonnes. As it is without convection currents, large temperature differences can develop between the places directly exposed to the Sun and the shaded regions. At night, the temperature plummets to $-170\,°C$, climbing again to $+110\,°C$ during the day.

The lunar day and night each last fourteen terrestrial days. In twenty-eight terrestrial days, the Moon spins once about its axis, but it also completes one revolution around Earth (which defines one 'month' on our satellite). This coincidence between the lunar day and a 'month' results from tidal coupling (see Fig. 1.1). Further consequences of tidal phenomena will frequently be referred to in what

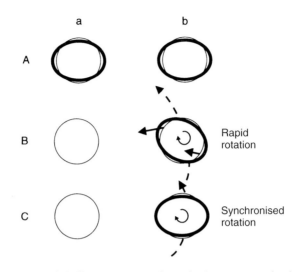

FIGURE 1.1   The tidal effect. (A) Two relatively close massive bodies, a and
b, such as Earth and the Moon, are deformed into ellipsoids by the gravitational forces they exert upon one another. To simplify, only the deformation of body b is shown in the diagrams B and C. (B) The two objects
rotate about their common centre of mass (only the orbit of b is shown),
and simultaneously spin on their own axes. If the objects spin with
shorter period than their orbital period, the major axis of the ellipsoid is
not instantaneously aligned with the axis of the system (a straight line
joining the centres of a and b). This is because body b cannot respond
infinitely quickly to gravitational deformation by body a. In this case,
the gravitational force tends to align ellipsoid and system axes, decelerating the spin of b about its axis. (C) Ellipsoid b slows down until the
rotational period about its own axis equals its orbital period (synchro-
nous rotation). From this point on, it always presents the same face to
body a, whilst continuing to spin on its own axis. This is the situation for
the Moon and Earth, and for most other moons in the Solar System. The
same process has slightly reduced Earth's speed of rotation about its
own axis. Four hundred million years ago, the day lasted only 21 hours.

follows. The Moon therefore rotates about Earth in such a way that it
always shows the same face. The first photos of the hidden face of the
Moon were taken by the Soviet probe Luna 3 in 1959.

  The lunar landscape was wonderfully summed up by Edwin
B. Aldrin, who accompanied Neil Armstrong on the Moon, when he

exclaimed: 'What magnificent desolation!'. In fact the noun describes the reality much better than the adjective. Without atmosphere, the lunar sky is always black, even when the Sun reaches its zenith. It cannot be compared with the beautiful blue sky here on our own planet, where molecules in the air tend to scatter those wavelengths of solar light which correspond to the colour blue. Under the dark lunar sky stretches a desert landscape, pitted with craters of all sizes. These may be as much as 300 kilometres across, but the smaller they are, the more frequently they occur. Despite the widely held view that they were due to volcanic activity, a hypothesis upheld until the middle of this century, they are actually the result of meteorite impacts. These vagabond rocks streak through interplanetary space at speeds of several tens of kilometres per second. From time to time, their path will intersect the orbit of some other body, such as the Moon or Earth. In the case of our own planet, these projectiles vaporise as they enter the upper layers of the atmosphere at supersonic speeds. Only the largest of them succeed in reaching the ground without being completely vaporised. Fortunately, such objects are extremely rare. Without any atmospheric shield, however, the Moon is constantly bombarded by meteorites. In the past, this bombardment was considerably greater. Hammered in this way by so many violent impacts, the lunar surface has gradually been transformed into a fine powder, known as the regolith, which forms a layer several metres thick.

Devoid of any atmosphere, the Moon has no magnetic field either. This has significant consequences. Indeed, interplanetary space is teeming with high-energy charged particles (electrons, protons and heavier nuclei). This is partly due to the solar wind, a flow of particles from the solar corona, and partly due to cosmic rays, particles accelerated by stellar explosions within our Galaxy. Earth's magnetic field protects us from these harmful particles, deflecting them towards its magnetic poles. In collisions with air molecules, they then lose their energy and it is the resulting radiation which gives rise to the spectacular northern lights, or aurora borealis. Lacking any atmosphere or magnetic field, the Moon is incessantly bombarded by these microscopic projectiles.

The lunar environment is hostile and manned missions expensive. This naturally raises the question of whether a further visit is justified. Apart from national prestige (which we might hope to be an obsolete motivation), and a few bits of stone, what could be the fruits of further manned lunar exploration?

An ideal observatory?

Astronomers are among the keenest advocates of a return to the Moon. In fact, our natural satellite is a genuine museum, keeping intact a complete record of its past. On Earth, such traces have long disappeared under effects of volcanic activity, tectonic motions of the plates making up Earth's crust, and erosion caused by water flow and atmospheric circulation. As Earth is still a living planet, all this activity tends to obscure any historical record. (Some information can be recovered, through fossils deposited in various geological strata.) In contrast, the Moon has long been geologically dead. Lava has not flowed on its surface since about 3 billion years ago, when the lunar crust completely solidified. As there has been no erosion, our satellite constitutes a kind of natural archive of all meteoritic impacts and all energetic particles it has ever intercepted. If these archives could be read, part of the history of the Solar System might then be reconstituted. In particular, it would be possible to establish the evolution of meteorite frequencies and solar wind intensity.

However, astronomers do not consider the Moon simply as a window on the past. Its various features make it an ideal site for astronomical observation. The fact that there is no atmosphere means that the whole spectrum of electromagnetic radiation would be accessible. This represents our main source of information concerning the rest of the Universe. The terrestrial atmosphere absorbs almost all of the electromagnetic spectrum, letting only visible and radio frequencies through to ground level. Other frequencies have only become accessible with the advent of the space age, when infrared, ultraviolet, X-ray and gamma-ray detectors could be carried aboard satellites.

A further disadvantage of Earth's atmosphere with regard to astronomical observation is turbulence. Incessant motions of its different layers cause a continual fluctuation in the refraction of light rays passing through. Stars twinkle and appear as fuzzy patches, even when viewed through the world's best telescopes (although recent sophisticated techniques, known as adaptive optics, can consistently improve the situation). Extremely clear optical images would be obtained on the Moon in the absence of these effects. In addition, continuous observation of optical frequencies would be possible, through day and night alike, under the ever-darkened lunar sky (except in directions close to the Sun and Earth). These frequencies cannot be observed from Earth during the day, or in cloudy conditions. The Moon's slow rotation would also allow the making of much longer exposure photographs than can be obtained on our own planet.

Weak gravity is another asset available for Moon-based astronomy. As telescopes grow larger, there comes a point where their own weight can cause mechanical deformations. Much bigger, and hence more powerful, telescopes could be built on the Moon. The lunar telescope would weigh six times less than its terrestrial counterpart of comparable dimensions. Constructing these lunar telescopes and thus eliminating the problem of atmospheric turbulence, images of distinctly higher quality could be produced than those from the Hubble Space Telescope.

Infrared observation would also benefit from the lunar environment. Infrared telescopes are sensitive to 'cold' objects, with temperatures a few tens or hundreds of degrees above absolute zero (-273 °C). Unfortunately, our planet has a mean temperature of 290 K (or $+17$ °C) and itself constitutes a strong infrared emitter. Setting up telescopes in the Antarctic or in space, this interference emission can be reduced somewhat, but never completely. At the South Pole, the temperature never falls below 180 K (-93 °C), whilst space telescopes spend a good part of their orbits bathed in solar light and heat. The infrared background is reduced by cooling with liquid nitrogen or helium when telescopes are based in the terrestrial neighbourhood. Needless to say, this greatly increases the complexity and cost of infrared missions. In contrast, the polar regions of the Moon shelter some of the coldest corners of the Solar System. As the Sun is always low on the horizon in these regions, its light merely skims the peaks which lie around the craters there. Conditions of permanent shade deep down inside certain craters guarantee temperatures below 50 K (-223 °C). These would be ideal sites for housing infrared telescopes.

Radioastronomers are also interested in the Moon, but for another reason. The problem on Earth is radio 'pollution', by signals of every kind (e.g., military radar, radio, television, satellite communications). The situation is likely to worsen, as telecommunications spread.

Despite increasing awareness of the threat amongst astronomers, who are attempting to limit this pollution in certain frequency bands, the trend does not appear to have been contained. The only place near Earth to escape interference from these terrestrial activities is the far face of the Moon. In addition, there are two powerful radio emitters in the Solar System, namely, the Sun and Jupiter, the largest planet. These two sources are not present in the sky of the Moon's hidden face for half and a quarter of the time, respectively. A radiotelescope on such a site could survey the sky in ideal conditions for six consecutive days in every lunar month, a situation radioastronomers can only dream about today!

However, not all features of the Moon represent advantages for astronomical observation. About a hundred microcraters (diameter greater than 0.05 mm) are formed each year by meteorite bombardment for every square metre of the Moon's surface. Telescope mirrors would have to be protected from this effect. In addition to this, dust raised by nearby meteorite falls (or even human activity in proximity) would gradually cover instruments and decrease their performance. Bombardment by high-energy particles would lead to breakdowns in electronic systems and extreme temperature variations between day and night would put equipment to a severe test. Composite materials with extremely small contraction and expansion under temperature change would have to be used for telescopes on the Moon.

But the greatest obstacle to lunar astronomy today is its high cost. Even on Earth, an Antarctic observatory costs between five and ten times more than its equivalent on the American continent. According to American astronomer Jack Burns, one of the best-informed and enthusiastic supporters of lunar observatories, such installations would cost between ten and a hundred times more than their terrestrial equivalent. Clearly, lunar telescope projects must be subjected to rigorous selection. Only exceptional results would justify the investments involved.

No date has yet been fixed for the first telescopes on the Moon. Very detailed mapping and better understanding of the lunar surface are essential prerequisites for such a project. The size of the first instruments would be limited by the high cost involved in transporting terrestrial materials. These instruments would be entirely automated and controlled from Earth. Once they had proven their reliability, more sophisticated instruments could be envisaged, which would be built on the Moon using lunar materials. As we shall see shortly, our satellite contains all the elements needed to build observing equipment, and even a manned base. In fact, it would be difficult to imagine running large telescopes on the Moon without setting up a permanent base. Human intervention seems unavoidable in building and maintaining complex instruments. It should be emphasised that exploiting the exceptional potential of lunar astronomy will far exceed the capacities of a single nation. These projects could only be achieved through cooperation on an international scale, and this not before twenty years from now.

Back to the Moon

For a long time only scientists showed any interest in returning to the Moon. Official enthusiasm, echoing the view of the general public, had rapidly died down after the Apollo programme. It was only in 1989, on the twentieth anniversary of the first manned lunar landing, that the United States once again officially manifested a desire to go back to the Moon. At the end of the cold war, American president George Bush considered that the scientific potential of his country could be freed for more pacific tasks, and launched his Space Exploration Initiative.

Led by Sally Ride, the first American woman in space, a group of scientists set out the four principal objectives of this ambitious pro-gramme: systematic monitoring of planet Earth, exploration of the Solar System by unmanned probes, sending astronauts to Mars, and a return to the Moon, but this time with a view to setting up a permanent lunar base.

It was intended that humans should return to the Moon in the first few years of the next century. The Space Station, described in one of the following sections, was assigned a key role in the project, namely that of a launch pad. Astronauts and material would be brought to the Space Station by a new heavy-duty launcher. They would then be transferred to lunar orbit by a special-purpose vehicle and lowered down to the surface by a lunar landing module. The task of the first crews would be to set up scientific instruments and the first modules of the lunar base. By 2005, the base would already be capable of housing five or six people for a period of several weeks. This initial capacity would gradually be increased to around thirty people by 2010.

Unfortunately, the United States Congress did not support their president in this initiative. It was judged too expensive for the targets envisaged. This failure did nothing to quench the enthusiasm of those who supported a return to the Moon. In the 1990s, new projects were devised in the United States, Japan and Europe. Although very different in approach and priorities, these projects did share one common feature: an effort to cut down considerably on costs. The Apollo programme had cost between 25 and 30 billion dollars at the time, equivalent to more than 100 billion dollars today. The Ride report had not estimated the cost of a return to the Moon, since it involved a vast programme spread out over several decades. However, the plans required construction of a space station, a heavy launcher and a transfer vehicle operating between the space station and lunar orbit. Costs would certainly have been much larger than that of Apollo. To put the figures into perspective, NASA works on an annual budget of 15 billion dollars.

In 1992, engineers at the American aeronautic company General Dynamics put forward a significantly reshaped version of the Ride project, which greatly reduced the cost of returning to the Moon. Their project cut out the Space Station and appealed to proven technology. An improved version of the Titan launcher would be used instead of the heavy launcher envisaged in the Ride project, whilst the vehicle proposed for transfer to lunar orbit was an improved version of the last stage of the Centaur rocket. The various components for the lunar mission were to be assembled in low Earth orbit, but without involving a space station. The only new feature was the lunar landing module, which was to be redesigned from scratch. It had to be lighter and perform better than the one used in Apollo. According to General Dynamics, a team could be sent to the Moon within a relatively short period (5 to 7 years) and for a cost of less than 10 billion dollars. However, these estimates did not arouse the interest of NASA officials. In their view, it would cost almost as much to make significant improvements in existing equipment as to build a new generation of vehicles.

The European (ESA) and Japanese (NASDA) space agencies are also interested in our satellite. However, their approach is different. Their projects are much more cautious and longer term than the one proposed by General Dynamics. A key feature in their conception is *in situ* exploitation of certain lunar materials. Avoiding transportation of these materials from Earth means considerably cutting down the cost of the whole venture.

Oxygen is undoubtedly the most significant of the materials earmarked for *in situ* production. It plays three crucial roles: chemical fuel for rockets, essential gas for breathing and as a constituent of water. When a manned module lands on the Moon, half its mass constitutes oxygen for the return journey. Using lunar oxygen, the mission load could be reduced by half. Or put another way, the payload could be doubled with the same launcher. It would also significantly diminish the mass of air transported from Earth for the crew's survival.

An initial stage is planned for the Japanese and European projects, which involves only unmanned probes. Their objective would be to make a detailed inventory of lunar resources and judiciously select sites where robots could be deployed. This first stage would last about ten years. This would be followed by a second stage in which the robots would extract oxygen and other important elements from the lunar surface. They would work partly in automatic mode and partly by remote control from Earth. This kind of operation from a distance is becoming more and more common today in hostile environments, with the development of electronics and robotics. It will be widely used on the Moon. Radio signals from Earth reach our satellite in slightly more than one second. The same cannot be said for Mars, at more than 3 light-minutes from Earth, where such procedures would be impossible.

Various methods for extracting lunar oxygen would be tested during this stage. Oxygen is an extremely reactive element. It occurs in the regolith combined with several metals in the form of oxides: silicon (almost 40% of the total), aluminium, titanium, magnesium, iron, calcium and so on. Several different methods of extraction have been considered, including chemical methods, electrolysis and pyrolysis. However, the advantages and disadvantages of each technique have not yet been established clearly enough to make a choice possible.

The aim of the second stage would be to gain familiarity with lunar techniques and apply them on a semi-industrial scale. It would indeed be an essential step before human beings could consider setting up a base. As more and more complex astronomical and industrial systems

FIGURE 1.2   NASA's probe Lunar Prospector in orbit around the Moon. In 1998, Lunar Prospector reported the presence of water in the lunar south pole. (Courtesy of NASA.)

are installed on the Moon, the number of astronauts must inevitably increase. This would constitute stage three. Its objective would be to develop a lunar base, sporadically occupied in the first instance, but then occupied on a more regular basis until permanent occupation became possible.

Lunar base

The lunar surface represents an exceptionally hostile environment for the life and work of human beings. Some tasks would be facilitated by the weaker gravity, but others, such as traction, become more difficult. Light contrasts between sunny and shaded regions may cause problems not only for human sight, but also for visualisation by certain automated systems. Light detection systems could be saturated by sudden exposure to sunlight. The almost perfect vacuum can modify properties of materials. For example, substances which would be in liquid form on Earth could vaporise on the Moon, whilst motor oil becomes a kind of viscous glue. Extreme temperature differences cause enormous strains in mechanical systems, especially those required to operate continuously. Dust is yet another complication in many tasks. More than half the mass of the lunar regolith is made up of ultrafine grains, almost invisible to the naked eye (dimensions less than one tenth of a millimetre). Such grains pass through the smallest space, getting into even the best protected joints and fixing onto any kind of surface by virtue of their static electricity. Protection and maintenance of equipment in these conditions demands a good deal more effort than it would on Earth.

However, the main danger for astronauts is the incessant bombardment of the lunar surface by solar wind particles and cosmic rays. American crews spent only a few hours on the Moon, corresponding to relatively low exposure. Shelter from these harmful particles would be a priority at the lunar base. Elaborate and costly constructions have been suggested, using a sort of concrete made from materials available at the lunar surface. Most projects involve covering living quarters with a layer of lunar regolith one or two metres thick. It has also been suggested that explosives could be used to enlarge holes made in the surface and thereby create a genuine underground base on several levels. NASA engineer Friedrich Hortz has put forward an idea to build the base inside caves, which may have been formed by lava flows on the Moon several billion years ago. In fact, photos of the surface show formations strongly reminiscent of canals. As liquid water has certainly never flowed across the surface of our satellite, these canals could only have been made by lava flow. Volcanically produced caves are known today on Earth, in places such as Hawaii. In fact, they are much longer than they are wide or high and look very much like tunnels. Some canals in photos of the lunar surface seem to disappear into the ground, implying that they may be volcanic caves whose ceilings have collapsed. If this turns out to be right, lunar caves may well provide a natural site for setting up base.

In the long term, a major objective of these projects would be to make the base as self-contained as possible. Unlike terrestrial installations in hostile surroundings, such as the Antarctic, transportation of materials from Earth will always represent a major component of the cost. As already mentioned, oxygen can be extracted from the lunar regolith. Water may exist in the form of ice, deep inside polar craters on the Moon. According to this hypothesis, which dates from 1961, such ice results from condensation of cometary water, built up from occasional collisions between comets and the Moon. The exceptionally cold conditions prevailing in the depths of these polar craters (never exceeding 50 K) mean that ice could be preserved indefinitely. In April 1994, the American probe Clementine flew by the lunar poles. The pictures transmitted by its radar suggest that there may be ice in craters at the southern lunar pole. The total quantity would amount to 100 000 tonnes. Four years later, in 1998, the US spacecraft Lunar Prospector found evidence for ice buried half a metre below ground level at the lunar poles. The total amount may be much higher than the Clementine estimates, being somewhere around six billion tonnes. If this interpretation of data from the two spacecraft proves correct, it would be sufficient to supply the first round of lunar bases. If not, hydrogen must be imported from Earth and combined with oxygen on site to produce water. Another possibility would be to extract hydrogen deposited on the Moon by the solar wind. One square kilometre of the Moon's surface intercepts about 100 g of material from our star each year. This uninterrupted flow of particles, consisting mainly of hydrogen, has been gathering in the regolith for billions of years now. It could be extracted by heating the lunar surface to temperatures of several hundred degrees.

In the long term, the lunar base would also have to be self-sufficient with regard to its energy consumption. Miniature nuclear reactors carried from Earth could supply enough energy for a certain time. Solar energy could also be used. It would not be difficult to set up giant solar panels, given the weak gravitational force, unlimited availability of space and vast supply of silicon. Indeed, silicon is the second most abundant element on the Moon after oxygen and could be used to build enormous reflecting surfaces. However, the system would not operate during the long lunar night (14 terrestrial days). In order to obtain continuous running, panels would have to be set up at the lunar poles, constantly bathed in sunlight. In the longer term, a suitable arrangement of panels could be distributed elsewhere over the surface, in such a way that at least part of the system would always be illumi-

FIGURE 1.3   An outpost on the Moon, with a rover on the left.

nated by the Sun. A satellite system would then relay energy to the various installations.

The site for a lunar base was a question discussed early on in the history of astronautics, even before it was firmly established that the trip could actually be accomplished. The American pioneer Robert Goddard addressed the question as early as 1920. In his view, the best site on the Moon would be its north or south pole. In the craters of these regions, crystallised water could be found and used as fuel (after separating out the hydrogen and oxygen). An energy production unit could be set up in a zone permanently exposed to the Sun's rays. Naturally, adequate protection would have to be arranged to ward off meteorites and he suggested placing instruments under cover of lunar rocks. Goddard's proposals are just as pertinent today. The only thing this visionary could not have predicted was the danger posed by high-energy particles. Although cosmic rays had been detected in 1912, by the Austrian scientist Victor Hess, their harmful effect on the human

FIGURE 1.4   A lunar base, with solar cells for energy production on the right.

body was only discovered much later, after the atomic explosions at Hiroshima and Nagasaki.

Recent progress in biotechnology gives reason to believe that a lunar base may eventually become self-sufficient in terms of food requirements. The aim would be to create an artificial biosphere, a completely enclosed space in which fauna and flora would coexist in an almost closed circuit. This idea has been borrowed from the terrestrial biosphere, a much larger natural region running on solar energy. Present experiments aim to create a closed circuit on a reduced scale. Plants which feed on animal waste and ground minerals become in their turn food for animals. Carbon dioxide produced by animal respiration is absorbed by plants to give back oxygen. This oxygen can then be used by animals to breathe, and so on. Finally, in an ideal system, used water is purified and reinjected into the system. These principles would apply

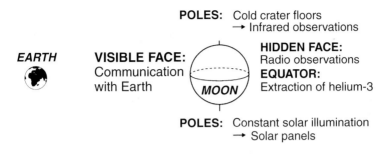

FIGURE 1.5   Sites on the Moon and their potential for installations and
lunar-based observation (see text).

not only to lunar bases, but to all future installations in space, particularly the more distant ones. In fact, as these techniques advance to the
point where they can ensure the autonomy of a certain habitat, it will
become feasible to set up more and more distant bases.

These projects no doubt fall within reach of the technological capabilities of our civilisation. However, they are likely to remain extremely
costly unless some revolution occurs in transportation technology. The
scientific interest of the Moon does not seem sufficient to justify such
costs.

Energy from space

One motivation for a return to our natural satellite could be our ever
increasing need for energy. Consumption today stands at around
10 terawatts (10 billion kilowatts) for a world population of 6 billion.
This corresponds to slightly less than 2 kilowatts *per capita* on average.
Consumption in developed industrial countries is between six and ten
times greater than it is in developing countries. Although it is difficult
to make reliable predictions, most views consider that the population
will rise steadily to reach 10 or 12 billion by the end of the next century.
This would imply a considerable increase in energy consumption,
brought on in particular by gradual industrialisation of developing
countries. Even if the industrialised countries were able to control their
own energy requirements, world consumption would very likely
exceed 25 terawatts during the second half of the next century.

According to present estimates, reserves of fossil fuels (e.g., coal,

FIGURE 1.6   A giant solar panel constructed in orbit around Earth.

petrol) will no doubt be sufficient for one or two centuries, and uranium reserves ought to last as long again. These estimates take into account not only the known deposits, but also speculated resources. Of course, past experience teaches us that such long-term extrapolations should be treated with caution. On the other hand, these resources are certainly non-renewable. Sooner or later they will be depleted. But even more worrying are the effects which accompany the use of such resources. Burning coal or petrol produces carbon dioxide which causes the greenhouse effect and contributes to global warming. In addition, the Chernobyl accident in 1986 and present-day problems of processing nuclear waste clearly emphasise the dangers and limitations involved in using nuclear energy.

Solar energy would appear to offer an attractive alternative. Each square metre of Earth's surface receives 1.4 kilowatts of this energy supply, which is virtually inexhaustible (at least, on a scale of several billion years). This almost equals today's average *per capita* consumption. After several years' research into solar energy, considerable

FIGURE 1.7   Constructing a solar power-cell in space.

progress has been made with regard to efficiency, ease of use and cost
of conversion into electrical energy. However, this renewable energy
supply could not cover all our energy requirements. Moreover, it is far
from equally distributed over the planet, favouring the lower latitudes,
and it is at best only available for half of the day. Another difficulty is
its extremely dilute form, requiring enormous collecting areas. As an
example, an area as large as Greater London would be needed to
provide for the energy consumption of a city like London.

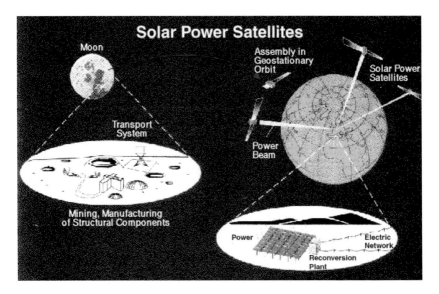

FIGURE 1.8   A scheme for solar power satellite construction and orbital transport. Lunar ore is processed and metals extracted for manufacturing structural materials on the Moon. They are subsequently transported in geostationary orbit, where solar power satellites are assembled. (Picture by Dasa-erno.)

At the turn of the century, Konstantin Tsiolkovsky had already suggested using solar energy as an electricity supply in space. Taking up the same idea, American physicist Peter Glaser observed in 1968 that very large solar panels could be deployed in space. The energy collected could be converted into a microwave beam directed towards a receiving antenna on Earth. The power station would be placed in a geostationary orbit at an altitude of 36 000 kilometres above the terrestrial equator. At this altitude, a satellite takes 24 hours to circle the Earth. This makes it permanently visible from the same point on the surface. At the end of the 1970s, NASA and the American Department of Energy studied the idea of a space-based solar power station. The basic project concerned a panel area of 54 square kilometres. This could collect 75 gigawatts (75 million kilowatts) of solar energy. Once converted into microwaves, the energy would be transmitted down to Earth by an antenna of diameter 1 kilometre built onto the power

FIGURE 1.9   A moon base (with a radiotelescope on the left) according to
plans of the Japanese Space Agency. (Courtesy of NASDA.)

station. Despite the very low spreading of the microwave beam, an
antenna of about 100 square kilometres would be needed to receive it
on Earth's surface. Global efficiency of all transmission and conversion operations would be lower than 10%, implying a final output of
only 5 gigawatts.

The power station would have had a mass of close to 50000 tonnes,
similar to the mass of a modern aircraft carrier. Getting all this
material into low Earth orbit (about 500 kilometres above the surface)
would have required about five hundred launches with a rocket like
Energia, the most powerful launcher today. A preliminary building
stage would be carried out at this altitude. The various modules would
then be propelled into geostationary orbit for final assembly.
Astronauts would certainly have been essential in controlling such a
complex series of operations, as complete automation was inconceivable in the 1970s. About sixty such units would be sufficient to cover
the needs of the whole of the United States. It would have cost something like 1 trillion dollars to realise the project, about ten times the

cost of Apollo. According to a NASA study, distributing this electricity over the national grid at the prices then in force, returns would barely have covered the cost of installation. The situation is slightly different today. Much lighter solar panels could be made than those available in the 1970s. Furthermore, developments in robotics would make complete automation a plausible option. Unfortunately, this kind of project is subject to a major drawback on strategic counts: all space installations are exceptionally vulnerable to missile destruction. Putting all energy eggs into one cosmic basket is tantamount to suicide. Such projects could only be envisaged under long-term guarantees of international peace and security.

Apart from the strategic aspect, building a space-based solar power station would still be an extremely heavy operation. The mere transport and assembly of such large quantities of material far exceed our present technological capabilities. One way round these difficulties has been put forward by NASA engineer Dave Criswell. His idea is to build the power station on the Moon, using materials naturally available on site. The advantages of the lunar surface for this kind of project are well known: low gravity, vast construction areas, and virtually unlimited availability of silicon and other metals. However, the drawbacks threaten to be at least as great. Continuous illumination could only be achieved by setting up a power station on either side of the Moon, or else installing one power station at one of the poles. The problem would then be to transmit the energy down to Earth. Our planet is never visible from the far side of the Moon, and is occasionally invisible from the poles, too. The latter is due to lunar libration, small oscillatory motions of our satellite about its axis of rotation. The energy gathered would then have to be transmitted via a relay satellite in lunar orbit. This would transfer the energy to a geostationary satellite, which would be the only way of continuously targeting the same site on the terrestrial surface. Needless to say, this spawning of intermediate stages would reduce overall efficiency and gathering areas would have to be made correspondingly larger. Moreover, the project would still necessitate very large solar panels in Earth orbit, to gather the solar energy transmitted from the lunar relay satellite, thereby thwarting the initial objective. This idea would cost much more than Glaser's projects.

Industrial opportunities on the Moon

Our planet's powerful gravitational field is at the root of the difficulties and high cost involved in transporting materials into space. A speed of 11.2 km/s, or 40000 km/hr, is needed to escape from Earth's gravitational hold. In contrast to this, a speed five times less, a mere 2.4 km/s, is required to escape from the Moon's weak gravitational field. The associated energy is proportional to the square of these speeds and is therefore roughly twenty times less for lunar escape.

The relative ease with which objects can be removed from the Moon's gravitational grasp suggests using lunar rather than terrestrial materials for space construction. This remark applies equally to projects in the neighbourhood of the Moon or Earth. Indeed, motion through space is a simple matter once outside the gravitational well of any heavenly body. The energy bill is almost negligible, unless speed is a key factor, in which case energy must be expended to produced the necessary accelerations. Once beyond the Moon's gravitational attraction, a speed of just 2 km/s more is sufficient to attain a geostationary orbit.

The absence of any atmosphere on the Moon also means that there is not the same friction with the air, which causes energy losses and heating during terrestrial launches. When leaving Earth, rockets only attain terminal velocity in the upper atmosphere, where the density is low enough to render friction negligible. Jules Verne's astronauts, launched by the celebrated Columbiad cannon in *From the Earth to the Moon*, would certainly have been roasted alive as they tumbled through the lower atmosphere at 11.2 km/s. The same risk is run on re-entering Earth's atmosphere. The astronauts aboard Apollo 13 narrowly escaped such a fate during their return in 1970.

One of the first to realise the importance of this problem was the English science fiction writer Arthur C. Clarke. In an article published in 1950, he proposed a device set up on the lunar surface which would catapult loads into space at a speed of 2.4 km/s. For several seconds, these loads undergo accelerations of about a hundred g (the acceleration due to gravity at the surface of the Earth). Naturally, the device was not intended for launching human beings. Our organism could not tolerate accelerations greater than a few g.

Clarke's catapult would look like a railway line, with two parallel rails several kilometres long lying across the lunar surface. Metal containers would be accelerated by a magnetic field, itself produced by powerful electromagnets set up along the path. At the end of the line, these containers would suddenly come to a halt, thereby projecting their contents into space. After two days' flight on a predetermined trajectory 65 000 kilometres long, the load would reach a temporary destination, namely a reception module strategically located at Lagrange point L2 of the Earth–Moon system.

In 1772, French mathematician Joseph Louis Lagrange made a study of two-body systems, such as the Earth–Moon system. He showed that in any such system, there are always five points where gravitational and centrifugal forces all cancel out. Three of these points are located on the system axis: L1 lies between the two objects, whilst L2 and L3 lie outside and rotate about the centre of gravity of the system with the same period as the two bodies. The other two points L4 and L5 form an equilateral triangle with the two bodies. Any object placed at these points will remain fixed in the same position relative to the two bodies (see Fig. 1.10). We shall meet these points again later on in this chapter. Point L2 on the Earth–Moon axis is located on a selenostationary orbit (from the Greek word *selene* for Moon). This means that it remains at constant distance from some given point on the lunar surface and can always be attained by the catapult, without having to alter the path of the projectile as the Moon or the reception satellite follow their own rotation. In the next stage, propelled materials would transit via this space freight system into an Earth orbit or some other region of the Earth–Moon system.

Clarke's electromagnetic catapult, or lunartron as it was known in the 1950s, provides an elegant solution to the problem of extracting objects from the Moon's attraction, and with low energy expenditure. A solar power station on the Moon, with a panel measuring 100 metres by 100 metres and having an efficiency of 30%, could supply enough energy to expel one kilogram per second. This amounts to about a hundred tonnes per terrestrial day, or a thousand tonnes per lunar day. At this rate, the volume of material required to build large space installations (such as a solar power station) could be put into orbit in less than a year. There appear to be no basic difficulties in the conception of the lunartron. Small experimental prototypes were even built in the

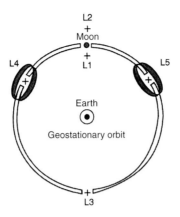

FIGURE 1.10   The five Lagrange points of the Earth–Moon system. The
geostationary orbit 36 000 kilometres above the terrestrial equator has
been drawn to scale. Points L1, L2 and L3 are located on the
Earth–Moon axis. They represent points of unstable equilibrium, in the
sense that any small perturbation would be sufficient to displace an
object lying there. Points L4 and L5 are equidistant from both the Moon
and Earth. They represent positions of stable equilibrium: an object
placed there tends to stay, even in the presence of small perturbations.

1970s, by Princeton physicist Gerald O'Neill. Loads underwent accelerations of several g and reached terminal speeds of 400 km/hr, one
twentieth of the lunar escape velocity. It should be emphasised that
extraordinary accuracy is essential for the trajectory of the catapulted
load. A target only a few dozen metres in diameter must be attained
from a distance of 65 000 kilometres. This is like aiming at a coin from
10 kilometres away. However, such precision is feasible today.

Assuming the electromagnetic catapult to be feasible, what would
be the most useful materials to extract from the lunar surface? The
obvious choice is oxygen, for the many reasons mentioned earlier, and
silicon, for construction of solar panels. Light metals such as titanium
and aluminium would also prove invaluable in building space installations. All these elements are abundantly available in the regolith, in the
form of oxides. However, they do not accumulate in seams like ore
deposits on Earth. Their low concentrations in the regolith would
make them difficult to extract.

In the long term, another element may turn out to be the most useful resource available on our satellite. This is helium-3 (^{3}He), a light isotope of helium which does not exist on Earth, but which may one day play an important role in solving our energy problems, as fuel for power stations based on thermonuclear fusion.

The idea of thermonuclear fusion is a simple one. Nuclei can be forced together by overcoming the mutual electrostatic repulsion between them. To achieve this end, they should have low electrical charge (implying light nuclei) and high speeds. The latter condition can be satisfied in the plasma state, at temperatures of several tens of millions of degrees. In thermonuclear bombs, such a temperature is obtained by first exploding a fission bomb. In stars, the only thermonuclear reactors known at present, it is gravitational attraction which confines the hot plasma and thereby maintains its high temperature.

The only way to confine plasma in a terrestrial fusion reactor is a powerful magnetic field. Any solid material in contact with the plasma would instantaneously vaporise. At the time of writing, plasma can only be contained near fusion temperatures for a tiny fraction of a second. The energy released during this brief instant is less than that expended to heat the plasma in the first place. Controlled fusion has still not been achieved despite forty years of research in several different countries. It is not yet possible to estimate a date for the first thermonuclear reactor, owing to the considerable obstacles which still block the way. However, based on the last four decades' experience, it is not likely to happen before at least two or three decades.

There are two reasons for pursuing this idea with such determination. The first is the cheapness of the fuel, deuterium (D), an isotope of hydrogen which is abundantly available on our planet. In fact, trillions of tonnes are held in Earth's oceans. (Reserves of hydrogen are tens of thousands of times more abundant, but the reaction between hydrogen nuclei is too slow to be of interest.) The second point is that reaction products are not radioactive, and thus do not pose the same serious problems as waste from nuclear fission. Operation of a fusion reactor nevertheless raises one rather serious difficulty. Conventional schemes for thermonuclear fusion at present being investigated in laboratory conditions involve the fusion of deuterium nuclei with other deuterium nuclei (or with tritium nuclei, another light isotope of hydrogen). About half the energy released is carried away by neutrons. These electrically neutral particles cannot be channelled by magnetic fields

FIGURE 1.11   A Moon installation with solar cells (in the middle), regolith processing facilities (in front) and a radiotelescope (on the rear). (Courtesy of NASA.)

and collide with the walls of the reactor, thereby rendering it radioactive in the long term. In contrast, the reaction $D + {}^3He$ produces few neutrons. Indeed, they carry away only 2 % of the total energy released, whence the interest in this isotope as part of a long-term energy programme.

Unfortunately, helium-3 does not occur naturally on Earth. Our planet's low gravity has allowed this light isotope to escape into space. Hydrogen and deuterium, although lighter than helium-3, have been retained on Earth as a result of their chemical reactivity, which has allowed them to combine with oxygen to form much heavier water molecules. Helium is barely reactive chemically and does not easily combine with other substances. Fortunately, the Sun contains considerable quantities of helium-3. In fact, about one nucleus of the solar wind in every hundred thousand is a helium-3 nucleus. The lunar surface has been bombarded for billions of years and large amounts of helium-3 have been buried in the regolith. Material brought back by Apollo astronauts and the Soviet Luna probes contains on average several micrograms (millionths of a gram) of helium-3 per kilogram. These materials were collected on the visible face of the Moon, which essentially only intercepts the solar wind when it happens to be facing the Sun. It is then lying in the tail of the terrestrial magnetosphere and is partially protected by Earth's magnetic field, which deflects charged particles in the solar wind. For this reason, the visible face of the Moon receives only about one third as many particles as the hidden face, which does not profit by such protection. In addition to this, the solar wind is more intense in the Moon's equatorial regions than in its polar regions (for the same reason that the Sun heats more at low terrestrial latitudes). The implication is that yields of helium-3 would be higher if extraction were organised on the equator of the hidden face (see Fig. 1.5).

Very large quantities of lunar regolith would have to be handled in order to extract significant amounts of this invaluable isotope, because of its low concentration in the lunar soil. As the solar wind does not implant it very deeply, vast areas of the Moon's surface would therefore have to be excavated. Almost a million tonnes of regolith would have to be processed, and this over an area of about 1 square kilometre, to obtain 10 kilograms of helium-3. Tens of thousands of square kilometres would have to be excavated to produce the hundreds of tonnes of helium-3 required for our energy purposes here on Earth. Such quantities could be despatched towards Earth by electromagnetic catapult or some more conventional means of transport. Although extremely costly, the operation would be profitable in the long term. In fact, the current price for the energy equivalent to the yield from 1 tonne of helium-3 (energy produced by other means today) is about 15 billion dollars. It is estimated that the Moon harbours enough helium-3 to supply our needs on Earth for the next couple of millennia.

On the other hand, if we succeed in controlling fusion reactions and if lunar helium-3 proves to be an economically viable fuel, the hidden face of the Moon is likely to become a vast scar under the destruction inflicted by our bulldozers. The science fiction story *The Lunatic Republic*, written by Compton Mackenzie at the beginning of the century, could serve to forewarn of this possibility. Arriving on the Moon at the end of the twentieth century, human visitors learn that its sterile, crater-studded surface is no natural accident. It is the work of 'intelligent' beings who inhabited the Moon long ago and whose miserable survivors now occupy a crater on the hidden face.

Economic activities in space

The 1960s space programmes had a merely symbolic value. They were intended to confirm the technological superiority of one nation over another. The last twenty years have profoundly changed this situation. Beyond scientific research and military applications, an increasing proportion of near-Earth space programmes now concern economic activities.

One category of these activities has already become part of our daily life: the ever growing transfer of information by telecommunications, navigation and terrestrial observation satellites. Remote sensing satellites generally occupy low orbits, at altitudes of a few hundred kilometres, encircling the planet several times a day. Many satellites fly in geostationary orbits, permanently stationed over the same region on the surface. It is often forgotten that Arthur C. Clarke advocated this type of orbit for telecommunications at the end of the 1940s. Today these orbits are so heavily used that there is real concern over long-term saturation.

In the future, there are two further categories of economic activity which may play an important role in our neighbourhood of space. The first concerns energy supply by solar power stations in geostationary orbit, described in a previous section. The second bears upon the manufacture of special-purpose materials whose properties may be difficult or impossible to produce on Earth. Space offers ideal conditions for this type of activity: an almost perfect vacuum and absence of gravitational effects, since all objects 'fall' in the same way in Earth's gravitational field. (Note, however, that motions of the capsule create microgravitational effects locally, although hundreds of thousands of times weaker than terrestrial gravity.) The vacuum is essential for the manufacture of extremely thin semiconducting films, used in computers among other things. Microgravity allows the formation of almost perfect crystals, of great importance in industry. In addition, elimination of convection in microgravity causes chemical fuels to burn differently. A study of combustion in these ideal conditions would help us to understand this complex process and improve the efficiency of motors back on Earth.

Many experiments of this type have been carried out in space over the past thirty years. However, no useful applications have yet been forthcoming and many industrial companies have withdrawn from this kind of research, despite their initial enthusiasm. At present these projects have no serious investment programme to back them up. High costs are of course the cause for this demise. Today, experiments in orbit cost between 20000 and 100000 dollars per kilogram of equipment, for unmanned and manned systems, respectively. The equipment may weigh several dozen kilograms and yield at most only a few grams of product. One kilogram of material produced in orbit would then cost about 1 million dollars!

The main obstacle to industrial development (or any other significant activity) in space is the high cost of getting into orbit in the first place. Present prices are prohibitive for any large-scale or long-term investors. But without private financing, it is hard to see how the necessary infrastructure could be set up, particularly as public funding is becoming scarce in the present economic climate.

According to some sources, the tourist industry may succeed in breaking this vicious circle. Even in 1967, the financial tycoon Hilton had considered extending his famous chain of hotels into near-Earth space by the end of the century. Soon afterwards, the airline company Pan Am studied the possibility of rocket-plane cruises in orbit around the planet.

Today, the Japanese are the only serious proponents of such projects. At the end of the 1980s, the world's largest manufacturer Shimizu revealed plans to construct the first hotel complex in space. It would have sixty-four rooms arranged in a ring 140 metres in diameter. The ring would spin about its axis three times a minute so as to create artificial gravity by centrifugal force. The whole complex would weigh about 7000 tonnes. Construction in low Earth orbit, planned for completion in 2030, would cost around thirty billion dollars. Among the activities envisaged were observation and photography of our planet, space walks, strange weightless sports near the ring axis and orbital weddings. It would surely be a simple matter to find a few thousand people willing to pay between 10 000 and 20 000 dollars for a night in such a dream hotel.

FIGURE 1.12   The international space station. (Courtesy of NASA.)

Although we are still a long way from realising these futuristic pro-jects, the imminent construction of the Space Station should provide a foretaste of what is in store. This NASA project, which has seen many alterations and changes of name over the past 12 years, started in October 1998. About fifty flights by the space shuttle and Russian, Japanese and European rockets will be needed to place all the neces-sary building materials into orbit. It will be the largest space construc-tion project yet attempted. It will weigh 400 tonnes and will be able to house a permanent team of six people in its aluminium modules (cyl-inders 9 metres long and 4 metres in diameter). Planned for completion in 2003, construction will cost around 50 billion dollars, as much as its running cost over the following ten years.

At the beginning of the twenty-first century, the international Space Station will certainly have become the most expensive thing ever built. This high cost, covered for the main part by the United States, has already provoked strong reaction across the Atlantic. Industrial and scientific benefits are considered by many as insufficient to justify such huge spending in a time of budgetary restrictions. For many others, however, construction of a permanent base in low orbit constitutes an essential and logical step in the conquest of space. Human and animal physiology could be better understood and it would be an opportunity for growing accustomed to working conditions in this hostile environment. All this is viewed as a ground-preparing stage for future developments, missions to the Moon and Mars, for which the Space Station could be considered as a launch pad. It reflects a long-term strategy, indeed maybe the only one which could justify its high cost.

O'Neill's space colonies

The idea that humans may one day settle in space, not only for work, but also to live permanently, is not a new one. Konstantin Tsiolkovsky had already considered the possibility at the turn of the century, followed by English physicist John Desmond Bernal thirty years later, and then many science fiction writers. These authors spent little time discussing the details of building and running space colonies, for they conceived such projects as being only in some indeterminate future.

At the beginning of the 1970s, American physicist Gerard O'Neill rediscovered these ideas and undertook a quantitative study with the help of his students at Princeton. The results convinced him that the project could become reality, and he presented it in his book *The High Frontier*. Published in 1977, the book was widely acclaimed by the American public.

In O'Neil's view, there were several reasons why humans would one day settle in space. Firstly, industrial activities could be concentrated there. This would avoid continuing pollution of the environment and exhaustion of terrestrial resources. Secondly, space would provide much-needed living space for Earth's ever increasing population. And finally, a new society could be founded in space, independent of terrestrial governments, thereby realising the age old utopian dream. Some find such reasoning rather naive. They argue that the same objectives could be achieved on Earth, by birth control, better management of industrial activities, and a general effort with regard to improving our society. However, experience shows that this programme may well be even more utopic than space settlement projects. We have already

FIGURE 1.13   Cylindrical space colonies according to G. O'Neill's plans in the 1970s. The three panels forming an angle with the cylinders reflect sunlight into the colonies through the transparent walls. Dozens of small agricultural colonies form a ring-like structure at each end of the cylinders. (Courtesy of NASA.)

proven that we can master the problem of travelling and working in space. But we have not yet proven that we can master the many facets of our own character.

The space colonisation project is based upon several simple observations. People need certain basic commodities in order to live: energy, air, water, food, a floor and gravity. The first of these is abundantly available in space, and the last can be simulated in any particular location. The others must be carried along. To lead a normal life, a person needs to have terrestrial weight, a 24 hour day–night cycle, natural light and an environment as close as possible to the one provided by the terrestrial biosphere.

It transpires that these conditions are best accomplished by means of a cylinder spinning on its axis. Living areas are arranged on the

FIGURE 1.14   The interior of a cylindrical space colony showing the three transparent walls alternating with residential areas. (Courtesy of NASA.)

inner walls of the cylinder, where centrifugal forces play the role of artificial gravity. The size of the cylinder depends on the number of inhabitants, and also on the building materials. O'Neill's projects concern populations of between ten thousand and several million people, in cylinders from 1 to 30 kilometres long (see Table 1.1). Just imagine how strange it would be for the first inhabitants of these cylinders as they raised their eyes to the sky and saw their neighbours' rooftops a few kilometres away, pointing in their direction!

O'Neill's cylinders rotate about their axes about once every minute in order to simulate terrestrial gravity on the inner walls. This period is too short to provide the right day–night cycle. That effect is obtained by a clever arrangement of three aluminium mirrors located outside the cylinder and rotating with it. The mirrors reflect solar light to the

Table 1.1. *Models for O'Neill's space colonies*

Model	Length (km)	Radius (km)	Population
1	1	0.1	10 000
2	3	0.3	30 000
3	10	1.0	100 000
4	32	3.2	1 000 000

colony via three glass openings situated along the cylinder. Each window faces a living area, so that the cylinder interior comprises three such areas separated by three transparent walls of the same size. The length of day is controlled by opening and closing the three mirrors. These fold down onto the transparent walls during the night. The same arrangement also determines the amount of solar energy received inside the cylinder, making it possible to control the average temperature and simulate seasons.

Solar panels deployed at one end of the cylinder supply the colony's energy needs. The axis of the system must point towards the Sun, to ensure that these panels are always facing our star. It is not easy to maintain an isolated cylinder in such a configuration. It would soon be destabilised by gravitational effects due to the Moon and Earth. O'Neill proposes a much more stable arrangement, produced by coupling two cylinders that rotate in opposite directions. This removes the need for stabilising rockets. The sets of mirrors create seasons in the two cylinders, one being in summer whilst the other is in winter. Their inhabitants can move easily from one to the other and enjoy a few days of holiday in a different climate.

An atmosphere, together with the thick cylinder walls, protect colonists from particles in the solar wind and cosmic rays. In the absence of earthquakes, volcanoes, floods and storms, the only type of natural disaster which can befall them is collision with an asteroid. These bodies cruise through space at speeds of several tens of kilometres a second. Fortunately, the probability of encountering such objects decreases rapidly with their mass. In the vicinity of Earth, an area of 1000 square kilometres intercepts one asteroid more massive than 1 tonne about once every million years. Projectiles more massive than about 10 grams could break the glass screens on the cylinders. These

FIGURE 1.15   The Stanford Torus was another popular design for space
colonies in the 1970s. (Courtesy of NASA.)

are more frequent. One per year could be expected to intercept an area
of 1000 square kilometres. The atmosphere's considerable mass would
prevent it from escaping very quickly through the broken glass and this
would allow time to make the necessary repairs. The problem is contained by continually monitoring the windows in such a way that the
slightest incident of this type can be spotted immediately.

Colonies obtain provisions from other cylinders nearby which are
devoted entirely to agriculture and stock farming. The farm cylinders
have weaker gravity and a warmer, wetter climate. In an entirely steril-
ised environment, plants, fruit and vegetables would grow without a
need for pesticides. Remote-controlled robots carry out most of the
work. By clever timing of the seasons in the various farm cylinders, colonists are assured fresh produce all year round.

O'Neill also speculated on the way people would live within this
futuristic society. The economy is based on manufacture of goods with

high added value (e.g., semiconductors, pharmaceutical products), which are then traded with Earth. No polluting forms of transport are used and the entire energy requirements are satisfied by solar panels. Industrial activities are carried out in special-purpose cylinders by remote-controlled robots. Cultural and leisure activities prevail in the daily life of space colonists. Near the cylinder axis, weak artificial gravity permits sports that would be unusual or impossible on Earth, such as flying some kind of vehicle (a flying bicycle!) merely through muscle power. More usefully, the old or handicapped can move around more easily in weaker gravity. This feature, combined with better climatic conditions and a purer atmosphere, make the space colony an ideal sanatorium.

O'Neill's space colonies would be set up at points L4 and L5 of the Earth–Moon system. These are equally accessible from both the Moon and Earth. Any object located near one of these points (within a radius of about 50000 kilometres) remains in stable orbit about it with the help of terrestrial and lunar gravitational forces. It is an ideal location for a large number of space colonies. If the first experimental attempts proved successful, the number and size of O'Neil cylinders would rapidly increase. The total population could attain several billion people.

Very large quantities of materials are needed to build space colonies. The mass of even the smallest cylinder exceeds half a million tonnes. Most of this is due to the frame and floor. It could not all be brought from Earth. O'Neill therefore suggested transporting it from the Moon by means of an electromagnetic catapult. He carried out an in-depth study of Clarke's lunartron, constructing several small-scale prototypes at Princeton. Moreover, he observed that such a scheme would be of no use in building models 3 and 4 (see Table 1.1), for they require tens of millions of tonnes of materials. In these cases, he suggested using materials from another distant, but not inaccessible site in the Solar System, namely the many asteroids which approach Earth from time to time. We shall return to this idea later in the chapter.

O'Neill evaluated the cost of the first colony in space at around 30 billion dollars (at 1972 values), comparable with the Apollo programme. Further colonies would cost less per unit mass, because a large part of the personnel and equipment would already be on site. According to O'Neill's original assessment, construction of the first

FIGURE 1.16   The interior of a Stanford Torus. Centrifugal forces create
artificial gravity on the outer walls, where residential areas are placed.
(Courtesy of NASA.)

colonies could have begun around the end of the 1980s, so that several
tens of thousands of people would already be living there by the beginning of the next century.

In 1985, O'Neill founded the Institute of Space Studies (ISS) in order
to promote and investigate the idea of space colonisation. His visions
have inspired many science fiction writers since they were so enthusias-
tically received in the 1970s. They also inspired the space city project
OLGA (Organic, Linear, Geosynchronous, Aesthetic), designed by
Italian architect Daniele Bedini in 1982. This is an ambitious futuristic
project founded on principles of architectural aesthetics and urban
planning. However, such ideas have little impact today on future space
projects. The conquest of space has turned out to be far more difficult
and expensive than had been imagined at the time. The old idea of
building bases on the surface of planets or their moons would appear

to be more realistic, at least for the coming century. However, there is no reason to exclude this type of project in the longer term.

A stairway to heaven

In the 1970s, rock group Led Zeppelin produced *A Stairway to Heaven*, which many consider to be the best piece of rock music ever written. The title reflects one of humanity's oldest dreams: to build a stairway up to the skies. This age-old fascination is well illustrated by biblical accounts of the Tower of Babel, or Jacob's ladder, which allowed angels to descend from heaven, according to the Book of Genesis. The Grimm brothers' *Jack and the Beanstalk* is a further demonstration of the same desire, in which the hero reaches the sky by means of a giant bean plant.

Today we have the means to reach the skies, or rather space, in the form of rockets. But what has become of our ancient dream? Can it be taken literally? It would be impossible to build a Tower of Babel right through the atmosphere, to an altitude of over a hundred kilometres. Beyond a certain height, any type of structure would collapse under its own weight. Employing exceptionally tough and light materials, such as carbon fibres (used among other things to make tennis rackets), it would in principle be possible to build a tower at most 40 kilometres high, nearly five times the height of Mount Everest. To support its own weight, the base of the tower would be about 6 kilometres wide, and it would become gradually narrower towards higher regions, exactly like the Eiffel tower. Carbon fibres are about as tough as the theories of materials science will allow. It is therefore clear that space cannot be reached by this approach.

In 1957, the Soviet engineer Yury Artsutanov realised that a cable transport system could, at least in theory, be set up between Earth and space, proceeding from top to bottom. The idea was to throw down 36000 kilometres of cable from a satellite in geostationary orbit and fix the end to the ground! As the satellite remains above the anchoring point, the cable would always be held perpendicular to the ground. Like a giant sword stuck into our planet, it would trace out a complete rotation every 24 hours. Courageous Jack in the Grimm brothers' fairy tale could climb up its 36000 kilometres and look down upon Earth from the middle of space. Artsutanov published his idea on 31 July

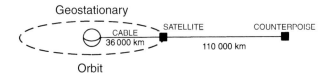

FIGURE 1.17 A schematic diagram of the space lift (not to scale). The weight of cable between Earth and the geostationary satellite pulls it down towards Earth. This force must be balanced by the centrifugal force exerted through the second cable (to a satellite as a counterweight).

1960 in an article aimed at young readers of *Komsomolskaia Pravda*, a communist youth newspaper in the former USSR. The same idea was independently discovered in 1966 by a group of American oceanographers, who are familiar with the laying of cables across the ocean floor, and then by the American air force engineer Jerome Pearson.

In Pearson's rather detailed plans, the cable was to be manufactured on board the satellite and lowered slowly down to Earth. The part of the cable to be fixed to the ground would be very fine. As work proceeded, the cable would get thicker, so as to support the increasing weight of its lower regions. Meanwhile, a second cable would be fed out of the satellite in the opposite direction. Its purpose would be to balance the satellite–cable system, fixing the centre of mass at 36 000 kilometres. The gravitational force on the descending cable would be balanced by the centrifugal force on the ascending cable. When the first cable reached the ground, the end of the second cable would be 110 000 kilometres above the geostationary satellite, reaching almost a third of the way to the Moon.

Lift cages carrying various loads would climb up the cable from Earth using electrical energy, at a cost of a few dollars per kilogram, compared with 10 000 dollars per kilogram in one of today's rockets. Once through the atmosphere, lifts could reach the geostationary satellite in about 6 hours, travelling at speeds of 6000 km/hr, comparable with today's fastest aeroplanes. Of course, friction with the cable would have to be avoided, by using a powerful magnetic field to levitate the cage. As the cabins gained height, it would become ever easier to free their contents from the hold of Earth's gravity. Beyond 25 000 kilometres, any object placed outside the lift cage would never fall back to

Earth, but would go into Earth orbit instead. Thrown into space at 36 000 kilometres, the object would barely move away from the geostationary satellite, but would merely follow it in more or less the same orbit. If the lift then continued right out to the end of the second cable, 146 000 kilometres from Earth, it could reach a speed of about 11 kilometres per second. Then, projected into interplanetary space by this tremendous sling, the object would finally escape from terrestrial attraction.

Electrical energy seems the best means of propelling the lift cages. But how could it be transported? The whole system could not be supplied by the kind of cables usually employed on Earth, for the energy losses would be enormous and the cables much too heavy. Energy would have to be produced locally, probably by means of solar panels located along the journey. An alternative would be to use the gravitational energy of cages on their way down, which would have to be decelerated anyway, to prevent them from crashing into the ground.

The advantages of this system are obvious. The cost of going into space is drastically reduced, and it removes the need for gigantic and dangerous rockets to snatch loads from Earth's gravitational attraction. On the other hand, quite phenomenal quantities of material would be needed to build the space lift. A carbon fibre cable capable of lifting a hundred tonne load up to the geostationary orbit would have a cross-section of a few millimetres at ground level and about ten centimetres near the satellite. Despite its lightness and ultrahigh performance, the cable would weigh over a million tonnes, more than the entire weight of some of O'Neill's space colonies! In fact, more than one cable would be required (one up and one down, both doubled for security reasons). In addition, the geostationary space station would have to be quite massive to keep the centre of mass of the whole system close to 36 000 kilometres above Earth. The total mass would be several million tonnes, and this would be difficult to transport from Earth, at least before having a lift available!

Arthur C. Clarke was the first science fiction writer to use the idea of a space lift in his novel *Fountains of Paradise*, published in 1979. The lower end of the cable touches the peak of a mountain in Sri Lanka, Clarke's adopted country for the past forty years. The cable is made from asteroidal material, following a suggestion by O'Neill some years before in the context of his space colonies. At the time, the toughest known form of carbon was diamond! Today, a third form of carbon is known: fullerene C_{60} (named after American architect Buckminster Fuller). A tubular form of fullerene, produced in an American laboratory in 1990, is even stronger than diamond, and it is probably the most suitable material for building the space lift. Clarke realised that it would be easier to apply the idea on Mars, its gravity being only 40% of that on Earth. Moreover, the areostationary orbit (equivalent to the geostationary orbit for Earth, from the Greek word *Ares* for Mars) is only 16 000 kilometres from the planet's surface. A Martian fullerene lift capable of transporting a hundred tonnes into space would weigh only a few thousand tonnes.

As Clarke emphasised, the many artificial satellites in orbit around Earth, not to mention asteroids, would constitute a considerable threat to such a lift. A collision at several kilometres per second with any of these objects could badly damage the cable or even break it completely. A few hours later, its lower section, now in free fall, would enter Earth's atmosphere at about 10 kilometres per second. The tremendous shock wave, carrying the same energy as several megatonnes of TNT, would propagate through the atmosphere along the cable's deadly path, as the cable wound itself around the equator like a giant snake. Just such a disaster is described in the science fiction novel *Red Mars*, published in 1991 by the American writer Kim S. Robinson. The author brings out an important point. Even if a natural disaster of this type could be avoided by destroying artificial satellites or asteroids as soon as a collision trajectory is confirmed, how could the cable be protected from deliberate sabotage? As we observed in our discussion of solar power stations, space installations are particularly vulnerable to missile attack.

Asteroidal resources

Since ancient times, and particularly since the industrial revolution, people have made increasingly intensive use of metals. The most common is iron. Annual production stands at around one billion tonnes today. Light metals such as aluminium and titanium, which are more and more widely used in industry, are produced in volumes one tenth and one hundredth of the quantity of iron, respectively. Generally speaking, the more abundant a metal is in Earth's crust, the more it is used and the cheaper it is to produce. Terrestrial metal deposits should cover the needs of our industrial civilisation for the next few centuries. Furthermore, metals can be recycled, unlike fossil fuels (coal, petrol and gas), which are destroyed when used. Metal shortages are therefore improbable, in either the near or distant future.

However, large amounts of energy are needed to extract metals from their ores (a tonne of coal is burnt for every tonne of iron produced). The ecological consequences are serious. In addition, deposits close to the surface are gradually being used up. This means that new sources must be sought deeper and deeper, thereby raising costs. For these reasons, it was suggested at the end of the 1970s that mankind should look to space for raw materials. Today it seems that such resources will only become viable for Earth-based consumption a few centuries from now. The great rush towards the new El Dorado will be motivated rather by the need for materials in large space installations, or even colonies. As observed in previous sections, the cost of bringing materials from Earth, even a few hundred kilometres in a straight line, are prohibitive. In contrast, relatively little energy is required to transport a massive object millions of kilometres across space, provided that no intense gravitational field is encountered on the way.

Solar energy has long been recognised as the source of all life on Earth. It is less well known that, since ancient times, people have used another resource from space. In about 3000 BC, the Hittites were the first to use iron from meteorites, stones fallen from heaven, to forge their swords. It was through the superiority of these weapons that this warlike people in the Middle East were able to vanquish the Egyptian army and conquer the realm of the pharaohs. The art of war was revolutionised in ancient times when iron replaced bronze in weapon production. The whole course of history was undoubtedly changed. It seems likely today that use of space resources will cause a similar revolution in the future history of space conquest.

Some claim that the spatial origin of meteoritic iron was known to the people of antiquity, because the Greek word for iron is *sideros*, very similar to the Latin word *sidus* for constellation. However, most linguists agree that the resemblance between the roots of these words is a mere accident. It is true to say that the phenomenon of shooting stars provides a natural link between the sky and iron meteorites found on the ground. On the other hand, Aristotle held that shooting stars were a purely atmospheric manifestation. In his theory, material ejected from volcanoes rose up into the atmosphere where it was set alight by contact with the first crystalline sphere surrounding our planet, the sphere of the Moon. In fact, in Aristotelian cosmology, the world beyond the lunar sphere was perfect and unalterable. No change was conceivable and hence no heavenly body could detach itself and fall upon our planet. Aristotle's thinking was so influential in the West that the extraterrestrial origin of meteorites was not recognised by the scientific community until the beginning of the nineteenth century. On 26 April 1803, a shower of meteorites rained down on the little village of L'Aigle in France. Accounts made by the villagers, together with the many fragments recovered, finally convinced famous physicist Jean-Baptiste Biot (hastily despatched to the scene), as well as the rest of the scientific community, that rocks really could fall from the heavens.

For two centuries, such rocks were the only sample of extraterrestrial matter available to us. More than three thousand have now been catalogued, and their fragments are exposed in museums the world over. It is estimated that about 10 000 tonnes of meteoritic material enters Earth's atmosphere each year, at speeds of the order of several tens of kilometres per second. Only the largest pieces reach the ground, having shed most of their mass by friction with the atmospheric layers. As we shall see in Chapter 3, some of these objects represent a real danger for life on Earth.

Meteorites which fall to Earth belong to a population of small bodies in the Solar System with trajectories lying very close to the orbit followed by Earth around the Sun. The first object of this type, asteroid 433 Eros, was only discovered in 1898. Today, more than 400 near-Earth objects have been listed. The largest, 1036 Ganymede, is about 40 kilometres across and has a mass of around 100 000 billion tonnes. The smallest, on the other hand, have linear dimensions less than 10 metres, and masses of only 1000 tonnes. The number of such bodies decreases rapidly with diameter. There are probably about 2000 objects having diameters greater than 1 kilometre, and over 70 000 with diameters ten times smaller. It is thought that there may be several tens of millions of these bodies taken altogether.

Near-Earth asteroids constitute only a small sample of much larger populations residing for the main part relatively far from our planet, in a zone enclosed between the orbits of Mars and Jupiter. Within the

Solar System it is standard practice to measure distances in terms of the astronomical unit (AU), which is the distance between Earth and the Sun, that is, roughly 150 million kilometres. The orbits of Mars and Jupiter lie at 1.5 AU and 5.2 AU, respectively, whilst the asteroid belt stretches from 2 to 4 AU. Most of these bodies have more-or-less circular orbits. However, from time to time, one of them may be deflected towards the inner Solar System by the perturbing effects of Jupiter's gravitational field. It will then follow an elliptical orbit, which may bring it fairly close to Earth. Calculations show that after several tens of millions of years, such bodies will be perturbed once again, this time by the gravitational fields of the inner planets (Earth, Mars and Venus), and ejected from these new orbits. Naturally, collisions sometimes occur with the inner planets or the Moon.

Most asteroids (about four out of every five) are composed primarily of silicon, iron and calcium oxides. This composition is like that of Earth's mantle. Almost one in five asteroids contains a significant quantity of carbon, water, hydrogen, nitrogen and other volatile substances. Finally, a small number, in fact about 3%, are almost exclusively made of iron, nickel and other heavy metals.

Among the various types of asteroid, the most interesting as regards our future space activities are those known as carbonaceous chondrites, which contain volatile substances. Water is clearly the key commodity. Essential for life and a first-rate solvent, it can also act as an absorbent shield against harmful high-energy particles streaming across interplanetary space. Water is easily transported in liquid form, or even in solid form, thus removing the need for a tank. In the cold of space, a block of ice can survive for long periods without significant losses. Apart from this, water decomposes into hydrogen and oxygen, two other extremely useful commodities. There is no substitute for oxygen in animal respiratory systems, nor as fuel for chemical rockets. Hydrogen is an effective fuel but risky to use, since it tends to burn explosively in an oxygen-rich atmosphere. Other volatile substances can also be used as fuel for space propulsion, or in the chemical industry. Finally, nitrogen would play the indispensable role of inert gas in the artificial atmosphere breathed by astronauts, just as it does for us here on Earth, and it would serve as a plant nutrient in the various artificial biospheres. Extracting these volatile substances from the asteroid would be a straightforward matter. It suffices to heat the regolith to a few hundred degrees. It is quite clear that asteroidal hydrogen, nitrogen and probably water, which do not occur on the Moon, would be of the utmost importance to lunar colonists. It is worth mentioning that in 1905 Robert Goddard, pioneer in the field of astronautics, had already suggested using oxygen and hydrogen contained in asteroidal water for propelling chemical rockets. It was only three quarters of a century later that his ideas were finally given serious consideration by NASA.

In principle, iron asteroids are the second most important class of space resource. They contain not only the familiar iron-group metals (nickel, chromium, manganese, zinc and cobalt), but also precious elements such as gold and platinum. The latter are extremely useful in various industrial applications, both on Earth and in space, thanks to their physical and chemical properties. For instance, gold is an excellent electrical conductor and also very resistant to corrosion, both desirable features in the production of electrical circuits. An iron asteroid of diameter 1 kilometre contains about 10 billion tonnes of iron, enough to cover our needs on Earth for about a dozen years. The nickel in such an asteroid, although only 10% of the quantity of iron, would last us a thousand years. The same asteroid also holds about 100000 tonnes of platinum and 10000 tonnes of gold. Its present market value would exceed 1000 billion dollars. According to present estimates, there are about a hundred nearby asteroids with these dimensions. Needless to say, if such large quantities of these metals ever reached the market, their prices would plummet.

These space El Dorados are not inaccessible, even today. In fact, about 20% of near-Earth asteroids would be easier to reach than the Moon. Compared with a similar lunar mission, less energy would be required to cover the intervening tens of millions of kilometres, to slow down and land on the surface, then take off and return to Earth. Using present technology, such a mission would take a few months.

Transporting the vast quantities of materials extracted from the asteroid back to Earth orbit would be a much more delicate problem. In fact, it would appear that the most economical option is to bring the whole asteroid back and carry out extraction closer to home. This idea can be attributed to Tsiolkovsky at the end of the last century. In his book *Dreams of the Earth and Sky*, the father of astronautics claimed that one day people would be able to guide asteroids as easily as riding

FIGURE 1.18 Asteroid mining. The asteroid is brought in geosynchronous orbit and the extracted materials are used for the construction of a solar panel. (Courtesy of the Space Science Institute.)

a horse, and hence benefit from their almost limitless resources. The idea was taken up once more in the 1960s, by the American engineer Dandridge Cole at General Electric. In his book *Islands in Space*, which came out in 1964, Cole suggested using nuclear explosions to deflect an asteroid from its path and guide it into near-Earth space. A few years later, the American astronomer Brian O'Leary had the idea of using the Clarke–O'Neill electromagnetic catapult to propel the asteroid. The device, set up on the asteroid surface, would use solar energy to eject part of the regolith into space. By Newton's law of action and reaction, the asteroid would be pushed in the opposite direction. The advantage of this method is that no material need be carried out to the asteroid as propellant.

The asteroid would travel for several years along a spiralling trajectory in the inner Solar System until it reached the terrestrial neighbourhood, herded along by its space cowboys. The astrominers could then begin their task, using solar energy to heat the surface of the object. A few hundred degrees Celsius would suffice to melt the regolith on rocky asteroids. Iron, nickel and other metals could be extracted by applying magnetic fields. Iron asteroids would be more difficult to deal with. They are much richer in metals, but also much harder. Temperatures of almost 2000 °C would be necessary to melt the surface, and this would allow alloys with unwanted properties to form. More sophisticated techniques must be found to handle this type of object.

It would be easy to deliver asteroidal products to space colonies located at Lagrange points, in geostationary orbit or even on the Moon. Transporting them down to Earth would be another matter. Some have proposed sending metals directly down to the surface in ten-tonne packets. Suitably streamlined, these packets would be decelerated as they passed through the atmosphere, before landing without significant ablation in desert regions of the Earth. However, this idea of sending artificial meteorites in the direction of our planet is somewhat worrying. In 1980, American astronomers Michael Gaffey and Thomas McCord suggested a less risky plan. This involved making the packets less dense than water by injecting gases into the molten metal during the extraction process. Such packets, suitably slowed down by the atmosphere, would be dropped right into the middle of the ocean, where they would float until recovered at a later stage (rather like American astronauts before the advent of the space shuttle).

During the second half of the next century, exploitation of nearby asteroid resources may well become one of the most important space-based activities. However, much remains to be learnt about their properties (e.g., composition, mass distribution, frequency) before the first missions of this kind can take place. There is another reason why we should improve our knowledge of these space vagabonds: the potential threat they pose to our planet, as we shall see in Chapter 3.

Martian chronicles

Mars will undoubtedly be the next step in the human cosmic adventure. The Red Planet already dominates space projects for the coming few decades. Although it is not our nearest neighbour, its orbit lying at least 30 million kilometres further than that of Venus, it is the only planet in the Solar System to provide physical conditions resembling those on Earth.

Public interest in the Red Planet dates from the end of the last century. In 1877, the American astronomer Asaph Hall announced the discovery of two small satellites around Mars. These were named Phobos and Deimos, meaning fear and terror, respectively, in ancient Greek, for they were the companions of Ares, the Greek god of war and counterpart of Mars. This was not a total surprise. The two Martian satellites had been mentioned in various works of literature long before their discovery, notably in *Gulliver's Travels* by Jonathan Swift and *Micromegas* by Voltaire! These premonitions were based on the influence of the great seventeenth century astronomer Johannes Kepler. He was quite convinced of the harmony of the spheres and, being a great mystic, believed that Mars must have two moons, because Earth had one and Jupiter was thought to have four (first detected by Galileo). Although the argument was false, the prediction turned out to be true!

Still in 1877, Italian astronomer Giovanni Schiaparelli announced even more sensational news: the surface of Mars was crisscrossed with what appeared to be canals. Actually, the translation *canals* for the Italian word *canali* suggested that they might be artificially constructed, something which Schiaparelli had not intended at all when announcing his discovery. Ironically, it was this confusion which stirred up such enthusiasm amongst astronomers the world over and stimulated construction of new observatories. An example is Camille Flammarion's Observatory at Juvisy-sur-Orge near Paris. Fifteen years later, American astronomer Percival Lowell confirmed Schiaparelli's observations using his brand-new telescope at Flagstaff in Arizona. In his view, the canals were certainly artificially made, an affirmation which caught the imagination of the general public. Soon afterwards the father of modern science fiction, Herbert G. Wells, published *War of the Worlds*. The novel describes a Martian invasion of Earth, in extremely convincing terms! The idea of an inhabited planet Mars was soon widespread.

However, Lowell's observations were not confirmed by his contemporaries. American astronomer Edward E. Barnard, working at the Lick Observatory in California, found no trace of the celebrated canals. Nor did the Greek–French astronomer, Eugène Antoniadi, using the largest refracting telescope in Europe at the Meudon Observatory in Paris. Moreover, in 1906, biologist Alfred R. Wallace (who cofounded the theory of evolution with Charles Darwin) showed that water could not exist in liquid form on the surface of Mars. This had no effect on the general enthusiasm of science fiction writers, who considered the Red Planet to be inhabitable for many years to come. Thus, between 1910 and 1920, Edgar R. Burroughs (creator of Tarzan) wrote a dozen stories about the fascinating and wild world of Barsoom, the name given to Mars by its inhabitants. His work, together with Ray Bradbury's much acclaimed series *The Martian Chronicles*, conspired to make the Red Planet extremely popular among science fiction readers. It was not until 1950 that a desert-like Mars with no trace of any canals appeared for the first time in this kind of literature, when Arthur C. Clarke wrote *The Sands of Mars*.

With the advent of space exploration, our knowledge of Mars has literally exploded in volume. Within just one decade, between 1960 and 1970, half a dozen American and Soviet probes were sent in the direction of Mars. Two of them, the American Viking 1 and 2, actually landed on the surface and carried out *in situ* analysis of the ground and atmosphere. Their results showed that, although Mars is far from being the kind of living world imagined by Burroughs and Bradbury, it is nevertheless not completely without interest. Having witnessed an eventful past, it is likely to open up fascinating perspectives for the future!

It takes Mars 687 days to complete its eccentric orbit around the Sun. The Martian year is therefore almost twice as long as our own. On the other hand, one Martian day is almost the same length as our own, since Mars spins once on its axis every 24 hours and 37 minutes. It has an area of 120 million square kilometres, comparable with the area of dry land on Earth. Mars has one tenth the mass and only half the radius of Earth. Gravity there is less than 40% of terrestrial gravity, so that an escape velocity of about 5 kilometres per second is enough to tear an object from its gravitational hold.

The Martian atmosphere is composed mainly of carbon dioxide and is about a hundred times less dense than the terrestrial atmosphere. It nevertheless provides a low level of protection against high-energy particles in cosmic rays. Far from the Sun and without the benefits of a dense atmosphere (which could maintain the surface temperature by the greenhouse effect), Mars is indeed a frozen planet. The mean temperature is $-55\,°C$, but large variations occur at different times of day, in different seasons, and at different latitudes. The daytime equatorial temperature is above 0°C in summer, but can fall to −80°C during the night.

The surface of Mars is a rocky desert, rather like the one in Arizona. The regolith is essentially made up of iron oxides, giving the planet its characteristic red colour. Fine dust is raised by dust storms which may last for months, and hang in the Martian atmosphere for long periods. This dust scatters mainly low frequencies in the visible spectrum, so that the Martian sky has a pale pink-orange colour.

Mars exhibits some of the most remarkable features of all the planets in the Solar System (see Fig. 1.20). Tharsis Tholus, located on the equator, covers an area as large as Africa, at a mean altitude of 10 km, making it almost three times as high as the Tibetan plateau. Lined up along the eastern side of Tharsis are three giant volcanoes named Ascraeus, Pavonis and Arsia, reaching up 17 kilometres above the plateau, twice as high as Mount Everest. These volcanoes, like all the others on Mars, have been inactive for hundreds of millions of years. The most impressive by far is Olympus Mons, to the east of Tharsis. It is 27 kilometres high and its base measures 600 kilometres across, making it the biggest mountain in the Solar System. When Tharsis was formed several billion years ago, this appears to have fractured the Martian crust, producing a gigantic scar still visible today: the canyon of Valles Marineris. Its walls are 10 kilometres high and 100 kilometres apart, and it follows the Martian equator for about 4000 kilometres, almost one fifth of the planet's circumference. In the southern hemisphere, the meteoritic crater Hellas is the largest in the Solar System, measuring 2000 kilometres across and 4 kilometres deep. Thousands of other craters of all sizes bear witness to the tremendous meteoritic bombardment suffered by the Red Planet over billions of years (just like the Moon and Mercury).

One of the most spectacular features on Mars is the polar caps, which are primarily composed of water ice and carbon dioxide frost. They greatly vary in extent from one season to another. They shrink as carbon dioxide frost sublimes in the summer heat, and expand back to their winter dimensions in the colder season, when carbon dioxide once more condenses out. The northern polar cap is about three times as large as its southern counterpart, measuring almost 1000 kilometres in diameter and 5 kilometres thick. Liquid water could not possibly

FIGURE 1.19 Mars. The canyon of Valles Marineris runs almost parallel
  to the equator of the Red Planet.

exist in the temperature and pressure conditions at present prevailing
on Mars. In fact, only minute traces of water occur in the Martian
atmosphere, making it much drier than any terrestrial desert.
However, it is clear that water did once flow on the surface, billions of
years ago, at a time when the Martian climate was much warmer. The
bottom of Valles Marineris and other canyons (resembling dried up
river beds), together with certain geological indications, would attest
to the presence of rivers, lakes, rains and floods in the history of Mars.
Gradual cooling has steadily altered this state of affairs. Part of the
liquid water was absorbed into the frozen regolith to form permafrost.

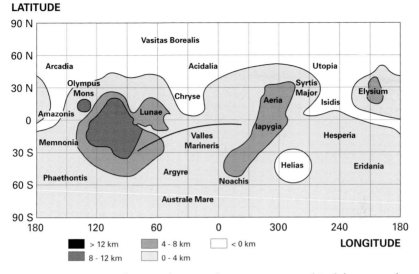

LATITUDE

FIGURE 1.20   Map of Mars showing the main topographical features of the planet (see text).

The remainder went into the polar caps or was trapped into meandering underground streams, rivers and lakes, several kilometres below the surface.

If water did once flow across the surface of the Red Planet, could it have engendered some form of Martian life? *In situ* investigations by the Viking landers in 1976 revealed no sign of biological activity, even at a microscopic level. This does not mean that there could be no life forms below the Martian surface, sheltered from cosmic rays and ultraviolet radiation. Twenty years after Viking, the question of life on Mars suddenly came back into the news. In 1996, spectacular revelations were made concerning meteorite ALH84001, which had fallen in the Antarctic 13 000 years ago. Analysis of its chemical composition suggested that the meteorite had Martian origins. Other results revealed the presence of a particular type of organic molecule, and even wormlike structures, which could be interpreted as signs of bacterial life. This conclusion was drawn by a group of NASA researchers and received as much media attention as Lowell's observations at the beginning of the century. However, it is widely contested today and is most likely false.

Apart from the all-important question of life on Mars, there are enough quite independent reasons for making the Red Planet our number one space objective for the coming century. It has already been the target of two further American missions: Mars Global Surveyor and Mars Pathfinder. In July 1997, the second of these landed a probe containing several experiments, and a small vehicle weighing about ten kilograms. The vehicle had six wheels and was able to travel a few hundred metres away from the probe, at a speed of half a metre per minute, carrying out analyses of the Martian regolith. In the coming decade, half a dozen other missions are programmed to place probes in orbit around Mars or on its surface, with a view to improving our knowledge of the planet. This should prepare the ground for the next stage: a manned mission to another planet in the Solar System.

Conquest of the Red Planet

German engineer Wernher von Braun, designer of the V2 rocket and father of the American space programme, was the first to carry out a technical study for manned travel to Mars. In his Mars Project, published in a German journal in 1952, he envisaged a fleet of ten spacecraft, each carrying seven astronauts. The craft were to be assembled in low Earth orbit by another fleet of space shuttles, which would also deliver the required 5 million tonnes of propellant. After a 9 month journey, fifty astronauts would go down to the Martian surface where they would remain for the next 15 months to explore the planet. The total length of the mission, including the return journey, would be three years, as long as the first trip around Earth by Magellan and his companions.

Von Braun estimated that the mission would cost about as much as a 'minor military operation', a colossal sum for any civilian operation. Ten years on, he put forward a simplified version of his project, requiring only two spacecraft and twelve astronauts. Half the crew of each vessel would go down to Mars, where they would spend 3 months exploring the planet. The cost of the operation would be considerably reduced by cutting its total length to 21 months and using a nuclear fission motor to propel the spacecraft. However, a special committee of the American senate put an end to all these projects in 1969. The Apollo epic was just being wound up and the time had come for budge-

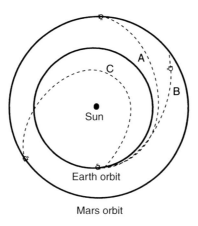

FIGURE 1.21   Trajectories of interplanetary spacecraft between Earth and Mars. Trajectory A is the Hohmann transfer orbit, along which the journey takes 260 days, and which involves a minimum of energy expenditure. Trajectory B is less economical, but requires only 180 days. These trips can take place when the two planets are in conjunction, i.e., the Sun and Mars are on the same side of the sky when viewed from Earth. The problem with this kind of path is that the crew must wait about 550 days on Mars before the time is right to get back (with the same outward and return journey time). The mission then lasts at least 900 days. Trajectory C applies when the two planets are in opposition, i.e., Mars and the Sun are diametrically opposed to one another with respect to the Earth. The actual travel time is much longer (about 430 days), but the stay on Mars can be reduced to just 30 days. (Adapted from Zubrin, R. & Wagner, R. (1996) *The Case for Mars*.)

tary cuts. Exploration of Mars by automated probes like Viking was the only option retained.

Using the means of propulsion available today, a trip out to Mars would take several months. In order to save fuel and thereby cut the cost of the mission, the vessel should follow an elliptical trajectory with one end touching the orbit of Earth and the other running into the orbit of Mars. In 1925, German engineer and architect Walter Hohmann observed that this type of path minimised energy expenditure when transferring an object between two planets. The problem with these Hohmann transfer orbits is that they take a long time to complete. Depending on the relative positions of the two planets at launch, a trip from Earth to Mars would take on average 9 months. Of course, faster journeys are possible, but they use more energy for the initial acceleration, and also for the final deceleration before an orbit around Mars can be adopted.

A space vessel launched into interplanetary space from Earth has a speed of 108 000 km/hr, the speed with which our planet orbits the Sun. It is this high speed (slightly increased by acceleration due to its own motors) which allows the vessel to travel the 400 million kilometres of the Hohmann orbit between Earth and Mars in just a few months. However, the Red Planet moves along its orbit at a slightly lower speed than Earth, 86 000 km/hr. The vessel therefore arrives with a relative velocity of about 20 000 km/hr, roughly equal to the difference between the two orbital velocities. In order to assume an orbit around Mars, it must decelerate to 2700 km/hr, otherwise the planet's gravitational attraction will not be sufficient to gain hold, and it will fly off into space. Such a deceleration, and possible landing, involve a great deal of energy consumption, as do the reverse operations required for the return journey.

Apart from the financial considerations, serious health problems are posed for the crew by the length of these missions. Experiments carried out by Soviet astronauts aboard the space station Mir have shown that long stays in weightless conditions lead to deterioration in bone and muscle tissues, as well as a weakening of the cardiovascular system. These effects, which disappear a few days or weeks after return to Earth, can be partially countered by an intensive programme of daily exercise. Given that astronauts ought to be operational as soon as they arrive on Mars, the spacecraft would probably have to be endowed with artificial gravity. Needless to say, this does not simplify the task of the mission designers.

A second risk arises from high-energy particles streaming through interplanetary space. Particles in the solar wind have energies less than 1 MeV. (The mega-electronvolt, or MeV is an energy unit typically used in nuclear physics.) These particles can be stopped by a few centimetres of shielding. However, the solar wind intensity can increase significantly during solar flares, eruptive events on the surface of our star which perturb the terrestrial magnetosphere and radio communications. In contrast, the flux of cosmic rays from our Galaxy is almost

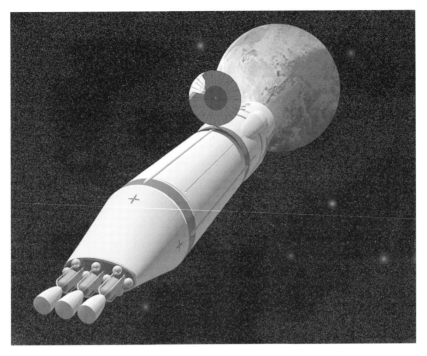

FIGURE 1.22 A spaceship approaching Mars. The radio antenna is pointing back towards Earth. (Courtesy of NASA/JSC, image # S97-0784.)

constant, whilst energies of individual particles are hundreds of times
greater. Several metres of shielding are required to block them out, and
this is not feasible for a spacecraft. Astronauts would have to face the
same problem during their stay on Mars. The Martian atmosphere is
thick enough to stop solar wind particles, but cannot keep out cosmic
ray particles. According to some estimates, the total exposure experienced by an astronaut during a two year trip to Mars (including return
journey and stay on the planet) would increase the risk of a fatal cancer
by several per cent. It should be emphasised that, although the risks of
exposure to an intense flux of particles are well known (following the
explosions at Hiroshima and Nagasaki), much less is known about
prolonged exposure to low-intensity fluxes.

In addition to the physical dangers which threaten the crews of
interplanetary missions, psychological difficulties should also be

considered. What would be the behaviour of a small group of people, forced to live and work together in an extremely confined space, through the long months of this dull journey? How would they feel, knowing that in case of difficulties, they could count only on their own resources (actually, almost non-existent), since Earth would be too far away? Such conditions represent a far greater constraint than those experienced in an Antarctic base, or by a nuclear submarine crew which might remain under water for months on end. In those cases, radio contact can quickly trigger a rescue operation. Even the astronauts engaged in lunar missions knew that, in case of difficulties, they could return to Earth within a few hours. Astronauts on the first missions to Mars must certainly possess the stuff of heroes.

Some have suggested that a mission to a nearby asteroid, lasting only a few weeks, would constitute an excellent training exercise for the crews of Mars missions. Others believe that practice in lunar space operations (at relatively modest expense) should precede any manned mission to Mars. In fact, the possibility of using lunar oxygen as rocket fuel, in combination with the weak gravitational effects on the Moon, make it an attractive launch pad for interplanetary missions. However, this cautious approach would delay manned missions to Mars by several decades.

The extreme difficulty of the project has not discouraged those who would like to see an assault on the Red Planet in the near future. A series of conferences under the heading *The Case for Mars* was organised in the United States during the 1980s. Their aim was to find technical solutions which could cut the costs involved in a manned mission. Indeed, the simplest schemes required quite exorbitant quantities of propellant (several hundred tonnes), which would somehow have to be transported from Earth's surface into low Earth orbit, in order to serve in the outward and homeward journeys, as well as during the Mars exploration itself. Even in the most optimistic view, such a mission would cost several hundred billion dollars, several times the bill for Apollo.

One way of drastically economising on propellant is air braking when the spacecraft falls through the Martian atmosphere. Friction against the upper layers of the atmosphere can produce considerable deceleration within the space of just two minutes. The craft then leaves the atmosphere with greatly reduced speed and goes into orbit around the planet. The brief period of braking would be a tremendous test for the astronauts on board. After several months in weightless conditions (or at best, low gravity), they would suddenly be crushed by the weight of their own bodies, reaching three or four tonnes during deceleration. The manoeuvre also entails another risk. The entry angle into the Martian atmosphere would have to be determined to an exceptionally high degree of accuracy. Any trajectory penetrating too deeply would cause the craft to vaporise by frictional heating, exactly as the smaller meteorites are destroyed upon entry into Earth's atmosphere. On the other hand, a trajectory with too great a tangential component would cause the vessel to bounce off the upper atmosphere and fly off into space, where it would be irretrievably lost. Whilst air braking is barely problematic in the case of unmanned probes, the same cannot be said when people's lives are at stake!

As for lunar missions, energy requirements can be drastically reduced by use of local resources. It has been suggested that the Martian moons, Phobos and Deimos, could serve as fuel supply stations. These bodies are in fact asteroids, measuring less than 30 and 15 kilometres across, respectively. Their weak gravity makes them easily accessible. Although their chemical composition is not well known, their dark colour and low density is reminiscent of the carbonaceous chondrites so rich in volatile elements. It may well be that the Martian satellites contain large quantities of water, either in the form of ice or absorbed into the regolith. Hydrogen and oxygen produced from this water could be used to fuel the descent to the Martian surface, and also the return journey to Earth. The idea was first put forward by Arthur C. Clarke in 1939, in the *Journal of the British Interplanetary Society*. Clearly, it was born before its day! Future unmanned missions should improve our knowledge concerning the composition of the two satellites. It is not yet absolutely certain that they contain water. However, introducing intermediate stages would greatly complicate logistics (and boost the cost) of this type of Mars mission.

Colonisation of Mars

At the beginning of the 1990s, American engineer Robert Zubrin suggested a simple and fairly realistic plan for a manned mission to Mars,

FIGURE 1.23 Propellant production from the atmosphere of Mars, according to NASA's Mars Reference Mission. Once the surface payload is unloaded, propellant production begins. The carbon dioxide atmosphere of Mars is reacted with hydrogen imported from Earth to make nearly 30 tonnes of oxygen and methane. These amounts are required to take a crew from the planetary surface into Mars orbit. The ascent vehicle is fully fuelled before the crew leaves Earth to begin the journey to Mars. (Courtesy of NASA/JSC, image # S93-050643.)

which served as a catalyst to NASA's own plans for the conquest of the Red Planet. This project, christened Mars Direct, is based on the use of Martian resources. A second advantage is that space vessels are launched from Earth with existing technology, and do not need to be assembled in orbit. The launchers proposed belong to the same category as the Apollo programme's Saturn V and the Russian Energia launcher. However, each mission requires two vessels, one manned, the other unmanned.

In Zubrin's plan, the first stage is to launch an unmanned spacecraft

FIGURE 1.24   Habitats and Rover on Mars. Once on Mars, the crew con-
nects the two habitats together and begins a variety of surface exploration and habitation activities. (Courtesy of NASA/JSC, image #
S93-45581.)

with a 30 tonne payload, containing the return module together with
necessary provisions, a mini nuclear fission reactor, a small chemical
factory, two exploration vehicles and 6 tonnes of liquid hydrogen.
Arriving on Mars eight months later, it decelerates by air braking and
lands on the surface. The nuclear reactor begins to supply the energy
needed to operate the factory. This absorbs carbon dioxide in the atmosphere and combines it with hydrogen to produce methane and oxygen.
Methane could be burnt as fuel to power Martian roving vehicles and
the return module. This approach was chosen because it is difficult to
store hydrogen, carried from Earth, in liquid form. The temperature on
Mars is well above the boiling point of hydrogen, so that expensive and
heavy cryogenic equipment would be needed. The factory also produces
an extra supply of oxygen by direct decomposition of carbon dioxide.

FIGURE 1.25   Mars ascent. After spending nearly 500 days on Mars, the
six crew members begin their 180-day voyage back to Earth by ascending into orbit to rendezvous with their Earth-return vehicle. Subsequent
human missions have the option of returning to the site established by
the first crew, or placing additional footholds on the surface of Mars.
(Courtesy of NASA/JSC, image # S93-50644.)

After 10 months' continuous operation, about 100 tonnes of
methane/oxygen, sufficient to cover the needs of the return journey, is
produced and stored. The control unit on Earth could then give the
green light for the next stage of the mission.

The manned spacecraft carries only four astronauts and their provisions for the outward journey (together with a third vehicle). Once its
fuel has been used up, the last stage of the launcher remains connected
to the manned module by a cable 1500 metres long. Small lateral
rockets set the whole system (manned module + cable + launcher
stage) into rotation about its centre of mass, with a period of 1 minute.
The centrifugal force creates an artificial gravity of 0.4g (40% of its
value on Earth) inside the module. In this way, the astronauts can grow

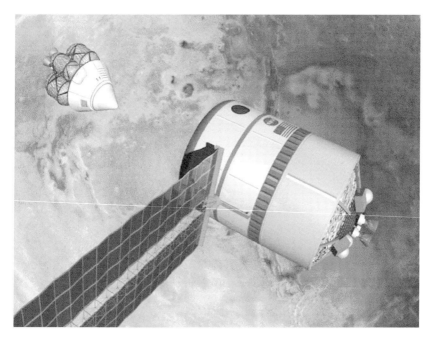

FIGURE 1.26   Rendezvous with Earth re-entry vehicle. The ascent vehicle
meets the Earth-return vehicle, which has awaited the crew's arrival in
Mars orbit for nearly three years. After checking out its systems, the
crew embarks on the final leg of their journey in the now familiar Mars
habitat. (Courtesy of NASA/JSC, image # S94-027626.)

used to gravitational conditions on Mars whilst they travel out to the
planet. On arrival, the cable is unhooked from the module, which
decelerates by air braking and drops down to the Martian surface, as
close as possible to the unmanned craft.

The crew stays almost 500 terrestrial days on Mars. This is because
Hohmann orbits require a certain alignment of Earth and Mars at the
time of launch. Favourable launch periods (or windows, as they are
known) only arise about once every two years. During their long stay,
the crew explores the surface of the planet with the help of three vehicles. All this time, they must wear pressurised suits, unless inside the
living module. When the next window occurs, the astronauts rendezvous with the return module and spend the next 6 months of the
Hohmann orbit in weightless conditions. Once back at Earth, they use

air braking in the terrestrial atmosphere to manoeuvre the module into low Earth orbit. They are then picked up by the space shuttle, after a total of 30 months on the mission.

According to Zubrin's estimates, Mars Direct would cost between 30 and 50 billion dollars; this would be devoted to development of the necessary equipment. This is just half the cost of Apollo and only one tenth the cost of classical projects for manned missions to Mars. A further 2 billion dollars would have to be added for each launch. Mars Direct plans to send spacecraft to Mars on a regular basis; in fact, every time a suitable window presents itself. In the long term, the aim is not only improved knowledge of the Red Planet, but also the installation of bases endowed with ever greater autonomy.

NASA further developed Zubrin's ideas and proposed its own plans, putting more emphasis on issues of crew safety and redundancy of various systems. In NASA's Reference Mission to Mars, five unmanned launches precede the departure of the crew, three of them in 2007 and two in 2009. The first puts the Earth-return vehicle into Martian orbit. It contains the re-entry capsule in which the crew will splash down into the ocean upon return to our planet. The second takes several pieces of equipment onto the Martian surface: the Mars ascent vehicle (to be used by the astronauts for their rendezvous with the Earth-return vehicle) as well as a propellant production module, a small nuclear reactor and a rover. The third cargo carries out a second nuclear reactor, a surface laboratory with food and supplies, and a remote control rover, all to be delivered to the surface of the Red Planet. Two years later, two more cargo launches put a second Earth-return vehicle into Martian orbit and a second ascent vehicle onto the surface. In this way all key modules (power plant, ascent vehicle, Earth return vehicle) exist in duplicate, to ensure redundancy in case any of them should fail.

It is only at that time, during the 2009 launch window, that the crew starts its journey on a 'fast transit' orbit that will bring them to Mars in six months. According to the plans, six or seven astronauts will spend almost 500 days on the surface of the Red Planet, taking the next giant step in mankind's space adventure. NASA's projects include a follow-up of several more missions to Mars in the years 2011–2014. Unforeseen developments (to be expected both on Mars and on Earth) make it difficult to establish a precise pattern for further steps in human exploration of the Red Planet. Still, it is conceivable that, after several such missions, significant amounts of equipment will accumulate on the landing site, allowing for the development of real Mars bases rather than just simple outposts.

The first step towards setting up self-sufficient bases would be to cultivate vegetable gardens under transparent plastic domes that could be blown up on site. These would absorb solar ultraviolet rays which are harmful to plants, and create a greenhouse effect that would warm up the regolith. Several different edible plants could be grown. They would be fed with atmospheric carbon dioxide and melted ground water. Unlike the Moon, Mars harbours all the volatile elements needed to create a biosphere. Of course, it could also supply all the metals required to build the various installations, including solar panels. Moreover, Mars may well be able to provide large amounts of areothermal energy, in a form which would be easy to exploit. According to present estimates, large fields of liquid water are located several kilometres below the surface, at temperatures of between 100 and 300 °C. Such depths could easily be attained on Mars by today's drilling techniques. Hot waters springing up from the Martian interior could turn the turbines of areothermal power stations, before condensing. Large quantities of water and cheap energy would certainly simplify the task of the first colonists. According to Zubrin's plans, several tens of thousands of people could be working on Mars by the end of the twenty-first century.

Although it seems feasible to set up Martian bases from a technical point of view, the length, risks and cost of the trip out to the Red Planet will all remain formidable for the next few decades. How then can we justify the cost of a Martian colonisation programme? Detailed exploration of the surface and study of its past history (in particular, with regard to the question of life forms) may not be sufficient motivation. Indeed, recent progress in robotics would indicate that exploration could be carried out equally well through an unmanned programme, at a fraction of the cost. Unlike the Moon, with its helium-3 resources and metal-rich asteroids, it seems that Mars has little to offer our Earth-based economy.

Supporters of manned Mars exploration recognise the importance of this point, but it does not mean that they have given up. In his recent book *The Case for Mars*, Zubrin emphasises the role that Mars could

FIGURE 1.27　The Mars base according to plans of the Japanese Space
　Agency. (Courtesy of NASDA.)

play as an outpost for missions to the asteroid belt. In a previous
section it was mentioned that near-Earth asteroids constitute only a
small fraction of the enormous population lying between the orbits of
Mars and Jupiter. This belt contains hundreds of billions of objects.
The largest is Ceres, 900 kilometres in diameter. The weak gravity on
these bodies makes access to their surface much easier. Like the nearby
asteroids, they are genuine open mines, representing considerable
potential for our civilisation over the centuries and millennia to come.

Because they are closer to hand, Martian bases could supply missions to the asteroid belt at much lower costs than Earth bases. Zubrin
argues that colonising the Red Planet would establish an economic triangle: Earth–asteroids–Mars. Earth and neighbouring space colonies
would supply Mars with high-technology products; Mars would
provide low-technology manufactured products and provisions for
operations in the asteroid belt; and finally, the asteroid belt would
supply metals to Earth and its space colonies.

This long-term perspective does not seem unrealistic. Several historical precedents show that it is often difficult to assess the economic potential of a distant territory. In 1803, Napoleon Bonaparte thus sold off almost one third of the present area of the United States for almost nothing (80 million dollars). In 1867, Tsar Alexander II sold Alaska to the Americans without ever dreaming of its gold and oil resources. In the same way, Martian resources may one day reveal quite unsuspected value.

Apart from its economic potential, Mars has already galvanised the imagination of scientists and science fiction readers, for it opens the way to a new world.

Terraforming

Humanity has sought to modify its environment from time immemorial. But actions were always restricted to a local level. Only God could create whole worlds, planets with a climate, flora and fauna. In his poem *Paradise Lost*, John Milton imagines God at work with his angels, tilting Earth's axis of rotation so that we might have seasons.

The idea that humans might globally modify a planet's climate appeared for the first time in fictional writing in 1930. In his major work *Last and First Men*, English writer Olaf Stapledon describes a vast project undertaken by our distant descendants to make the surface of Venus inhabitable. At the time almost nothing was known about atmospheric conditions on our sister planet, which lay unseen beneath thick clouds. Stapledon imagined seeding the surface with special 'biologically produced' plants, which would release the oxygen required for breathing by photosynthesis.

Twelve years later, American writer Jack Williamson invented a new word 'terraforming', in his short story *Collision Orbit*. The term meant transforming the surface of a body into an inhabitable world. Although Williamson's story was totally unrealistic (unlike Stapledon's), the terminology has nevertheless become widely accepted. However, some prefer to use the generic term 'planetary engineering', which covers any global attempt to modify the climate of a planet, reserving the word 'terraforming' for the creation of an Earth-like environment.

The first scientific article on terraforming was published in the journal *Science* in 1961 by American astronomer Carl Sagan. He discussed the possibility of making Venus into an inhabitable planet using micro-organisms. These would metabolise carbon dioxide and water (suspected at the time to be present in the Venusian atmosphere), fixing the carbon and giving back oxygen. The greenhouse effect would be weakened by reducing the amount of carbon dioxide, and this in turn would cool the Venusian atmosphere. Its temperature would fall well below its present level of about 700 K. As we shall see later in the chapter, this idea turned out to be unrealistic. However, it did inspire many science fiction writers, who went the rounds terraforming almost all the bodies in the Solar System!

Before describing a few recent terraforming projects, let us just recall that life has already modified the global environment of our planet. Several billion years ago, it was photosynthesis by blue-green algae, followed later by plants, which gradually raised the oxygen content of our atmosphere and made it possible for animal life to appear. More recently, mankind has unintentionally altered the climate of our planet by producing greenhouse gases (carbon dioxide and chlorofluorocarbons, or CFCs). Although our control over this situation leaves much to be desired, we have certainly proven our capacity to transform, for better or for worse, the atmospheric conditions on other planets.

Studies carried out since the 1960s have demonstrated that, of all the bodies in the Solar System, Mars is very likely the easiest (or rather, the least difficult) to terraform. Today, the Red Planet is hostile to any form of terrestrial life. Its atmosphere is too thin, too cold and unbreathable, whilst its surface is under constant bombardment by ultraviolet and cosmic rays. To move around on Mars, people would need heated, pressurised and oxygenated suits, just as on the Moon.

Water has nevertheless flowed on the Martian surface, proving that the atmospheric temperature and pressure were higher in the past. The planet could be returned to this state simply by providing it with a dense atmosphere, containing sufficient quantities of greenhouse gases. The average temperature would then increase and liquid water could once more exist on the surface. The pressurised and heated suit would no longer be necessary. Before the oxygen mask could be left aside, a great deal of oxygen would have to be introduced into the Martian atmosphere.

Today, Mars possesses all the volatile elements needed to make it inhabitable: water, nitrogen, carbon and oxygen (the last two being combined into carbon dioxide). However, they do not occur in gaseous form, but condensed into the ground and polar caps. The atmosphere contains very little carbon dioxide, producing a negligible greenhouse effect. Not enough is known about the Martian crust to be sure that reserves of these volatile elements would be sufficient to completely terraform the planet. Certain scenarios therefore appeal to extra-Martian resources. For example, they envisage introducing large amounts of volatile elements into the atmosphere by crashing a suitably composed asteroid onto its surface.

The number of studies devoted to transforming the Martian climate is impressive. They all advocate an initial warming of the atmosphere. As the temperature moves past certain thresholds, snowball effects are triggered. Through self-sustaining phenomena, the heating process is enhanced and there is no further need for human intervention.

The most detailed study to date is probably the one by the NASA research team Chris McKay, Owen Toon and James Kasting, published in *Nature* in 1991. Their favoured scenario begins by pouring huge quantities of CFC into the atmosphere, to boost the mean temperature by about twenty degrees. Several tens of billions of tonnes of CFC would be necessary, which could not be transported from Earth. A large number of CFC production units would be set up on the surface of Mars. To meet requirements for the project, annual production would have to be hundreds of thousands of times greater than the present level of terrestrial production. When the mean temperature of the planet reaches about $-35\,°C$, the polar caps and regolith begin to release their carbon dioxide content, intensifying the greenhouse effect and warming the planet up. However, it is believed that the total carbon dioxide reserves are not sufficient to reach a situation where the permafrost can thaw out and release liquid water. Some additional means must be brought into play. Carl Sagan and his colleague James Pollack have suggested using bacteria. These could survive on the planet as soon as the atmospheric pressure reached one tenth its value on Earth. They would metabolise nitrogen in the regolith to produce ammonia, which is another greenhouse gas. The temperature would then increase still further.

The combined action of several related processes (industrial CFC production, release of carbon dioxide, introduction of bacteria) would probably be required to take the first step in terraforming Mars. The planet would soon find itself surrounded by an atmosphere almost as thick as our own. However, it would be unbreathable, being composed mainly of carbon dioxide. Its mean temperature would hover around 0 °C, and this would thaw out the upper layers of permafrost in equatorial regions. According to estimates by McKay and team, this first stage would last at least one hundred years. The energy needed for this heating (to supply CFC production units and other installations) is equivalent to the amount of solar energy received by Mars over a period of about ten years. This should give an idea of the cost of the operation, which would exceed several hundred billion dollars.

At the end of this stage, the Red Planet would still be an arid desert. Although providing habitat for some bacteria and primitive plants, there would be no animals, nor even any highly developed plants. Indeed, the latter need traces of oxygen in the atmosphere to maintain their metabolism. Human beings could stroll around on the surface, provided they were warmly dressed and had an oxygen mask.

The following stage in terraforming Mars would be much longer and trickier than the first. The planet's *hydrosphere,* the system whereby water circulates between the regolith, the surface and the atmosphere, would have to be activated. The oxygen content of the atmosphere would also have to be raised. Natural mechanisms, such as photosynthesis by primitive plants, are too slow. Robert Zubrin and Chris McKay have thus revived an old idea for terraforming Mars which appeals to brute force. An enormous panel, a hundred kilometres in diameter, would be deployed in orbit around Mars to capture several tens of terawatts of solar energy. This energy would be focussed on the polar caps by means of an antenna, with a view to releasing some of the 5 trillion tonnes of water ice they are estimated to contain. Water would evaporate into the atmosphere, adding to the greenhouse effect and increasing the temperature. Much greater quantities again would be freed from the permafrost, as it thawed out to depths of several tens of metres. When this water condensed in the cold atmosphere, rains would fall on the Red Planet, after several billion years of drought.

As time went on, the dried-up canyons, river beds and underground cavities would begin to fill with water. Streams, rivers, lakes, seas and oceans would at last appear on the planet. Water would flow down the Valles Marineris, turning it into the largest river on Mars. A large part

FIGURE 1.28   Mars terraformed (painting by Michel Caroll.)

of the northern hemisphere, lying two to three kilometres below the average Martian ground level, would be covered with water. A continent as large as the Antarctic would emerge at the north pole, surrounded by the boreal ocean, almost as great as the Indian ocean on Earth. In contrast, there would be no oceans in the southern hemisphere. Several large craters such as Hellas or Argyre would become seas, similar to the Mediterranean in depth and extent.

Centuries on, the Red Planet would have lost its original colour, becoming blue and green. Indeed, over this period, solar energy would be used to heat oxides in the Martian regolith, releasing low levels of oxygen into the atmosphere. More developed plants could then grow on Mars, sustained by activation of the hydrosphere and propagating rapidly. These plants photosynthesise far more efficiently than primitive plants, and would gradually raise oxygen levels in the Martian atmosphere. Estimates by the McKay group suggest that many centuries would go by before the atmosphere could support animal respiratory systems. Particularly efficient photosynthesisers could be developed by genetic manipulation, in order to reduce the length of this stage to less than a thousand years.

The great length of the process is not the only inherent difficulty. The composition of the atmosphere would also have to be carefully adjusted. For example, oxygen levels should not be allowed to exceed a certain limit, which would render the atmosphere inflammable; and carbon dioxide levels must be kept down so that breathing remains possible. But if plant photosynthesis caused the carbon dioxide level to fall too low, it would have to be replaced by other greenhouse gases, to avoid renewed cooling of the planet. CFC production would have to be maintained on a permanent basis. Moreover, huge amounts of some neutral gas like nitrogen would have to be injected into the atmosphere, as Martian reserves of nitrogen are not sufficient. The volatile resources of carbonaceous chondrites would probably be required.

Creating new worlds

Although it would appear an extremely difficult task at the present time, it does seem that we might one day be able to make Mars inhabitable. Increasing its atmospheric density and temperature are apparently the least difficult aspects of the problem, whilst raising oxygen levels and activating the hydrosphere represent a major challenge. If planetary engineers succeed in meeting this challenge, our Solar System may possess two living worlds in coming centuries or millennia.

Venus is the only other planet in the Solar System to have aroused the imagination of world builders. It was soon clear that the task would be incomparably more difficult than for the planet Mars. Venus has a carbon dioxide atmosphere like Mars, but the similarity ends there. The Venusian atmosphere is much denser than our own, and its pressure is about a hundred times greater. Such a huge amount of carbon dioxide produces a tremendous greenhouse effect. The ground temperature on Venus is maintained at around 450°C (730K). Terraforming Venus would involve thinning out this smothering atmosphere, and thereby cooling the planet, quite the opposite process to the one envisaged on Mars.

In his original 1961 article, Carl Sagan had suggested using micro-organisms to fix carbon in the carbon dioxide gas and restore oxygen (respiratory gas without greenhouse effects). If this succeeded, the surface of Venus would be covered by a layer of graphite several hundred metres thick, lying under an atmosphere of pure oxygen. Such a transformation would not render the planet any more hospitable, because its atmosphere would be just as dense as before and would crush all living beings, apart from the micro-organisms. Furthermore, it would now be extremely inflammable. The graphite layer would burn very easily to form carbon dioxide once again.

Some have suggested blowing some of the Venusian atmosphere away by sending asteroids crashing onto the surface. Calculations show that an impact from a large asteroid (at least 700 kilometres in diame-ter) could expel almost one thousandth of the mass of the planet's atmosphere into space. In order to blow away a major part of the atmosphere, several thousand objects of this size would be required, whereas only a few exist in the Solar System. In 1989, Princeton physi-cist Freeman Dyson suggested using an enormous reflecting panel between the Sun and Venus, to plunge the planet into darkness and cold. The panel would be located at the Lagrange point L_1 of the Sun–Venus system. It would have to measure ten times the diameter of the planet to entirely shade it. Deprived of solar energy, Venus would cool very quickly. Carbon dioxide gas would condense out of the atmosphere to form an ocean on the surface. However, the rest of the process is much more difficult. Venus must be provided with a hydro-sphere (by importing water from one of the icy moons of the giant planets). Oxygen must be released from a small fraction of the carbon dioxide, by photosynthesising micro-organisms, for example, but on no account must the carbon dioxide oceans be allowed to evaporate back into the atmosphere. In other words they must be incorporated into the surface by chemical reactions. Needless to say, just the first step here, building a reflective panel 50000 kilometres across in space, will far exceed the technological and financial capacities of our civilisation for many centuries to come.

Terraforming Venus may prove to be one of the great challenges engineers will face in the third millenium. But regarding the other planets of the Solar System, there is today no possibility whatsoever (even on paper) of making them inhabitable. Mercury (the planet closest to the Sun) and Pluto (the most distant planet) are too small to retain an atmosphere. As far as the giant planets (Jupiter, Saturn, Uranus and Neptune) are concerned, they are composed of light elements, hydrogen and helium, which are useless or harmful to life.

The only other object to attract any attention in this respect is Titan, the largest moon of Saturn. It measures 2600 kilometres across and can be considered as a small planet, lying somewhere between the sizes of Mercury and Mars. On 12 November 1980, the American probe Voyager flew by the surface of this body, at a distance of only one Earth radius, or 6500 kilometres. Data returned to Earth revealed an exceptionally interesting world. Apart from Earth, Titan is the only other object in the Solar System to possess a nitrogen-rich atmosphere. The pressure at the surface is about 50% greater than it is on the surface of Earth, but the temperature rarely exceeds 95 K ($-180\,°C$). The Sun's faint light, a hundred times less intense than on Earth, barely filters through the thick fog of the atmosphere to illuminate the sinister landscape. The solid surface, composed mainly of ice, rock and solid methane, emerges from a viscous ocean of liquid methane. Indeed in the physical conditions prevailing on Titan, methane can actually exist in solid, liquid and gaseous forms. There may well be a methane cycle, just as there is a water cycle on Earth, with methane rains falling from its thick atmosphere. The Cassini probe, which was launched in October 1997 and is expected to arrive on Titan in 2004, should greatly improve our knowledge of Saturn's largest moon.

In the mid 1990s, Carl Sagan and James Pollack studied possibilities for modifying Titan's climate. The problem is quite different to those faced in terraforming Mars or Venus, because the pressure is already almost normal, and the atmosphere is composed mainly of nitrogen (like our own). However, low temperatures mean that the usual green-

FIGURE 1.29   In this artist's rendering, the European-provided Huygens
probe parachutes to the surface of Saturn's moon Titan after being
released by the Cassini orbiter. Cassini will receive and relay to Earth
scientific data collected by instruments on the probe, then explore
Saturn, its many moons, ring system and magnetic environment for
nearly four years. Cassini is a joint mission of NASA, the European
Space Agency and the Italian Space Agency and will reach Titan by
2004. (Painted for NASA by Craig Attebery.)

house effect substances (e.g., carbon dioxide, ammonia and water
vapour) are in solid form. The temperature would have to be increased
considerably before these substances could exist abundantly in
gaseous form. Sagan and Pollack suggest using thermonuclear fusion
of deuterium, a common element on Titan, to create a significant
initial heating effect. Once vaporised into the atmosphere, greenhouse
gases would take over the task of raising the temperature. According to
Sagan and Pollack's estimates, it would not be more difficult to heat up
Titan than to do the same for Mars. However, even if the temperature
climbed above 0 °C, sunlight intensities would probably be inadequate

for plant photosynthesis. It would be a hard task to oxygenate the atmosphere of Titan, a fact which seriously compromises the idea of interfering with its climate.

Not only science fiction readers, but also quite a few scientists are dreaming of ways to bring life to other planets in the Solar System, and in particular to Mars. Their motivation is certainly not a solution to overpopulation problems on Earth. Even though Mars has an area equal to all the land area on Earth, it would be impossible to transport any significant fraction of the population. In order to send a hundred million people (which constitutes a negligible fraction of the present population), in let us say one century, three thousand departures would have to be organised each day. Therefore, the fascination for terraforming Mars is more closely related to the new frontier it represents. Conquest of such a frontier would help our civilisation to release its creative potential and find new vitality. Some have compared the situation with the American frontier, several centuries ago. Conquest of America had many beneficial effects for the old continent (although unfortunately it led to large scale slavery and to genocide for the native peoples). Far from any established institutions, workers in the new world laid the foundation for a more democratic society, and this helped (along with other factors) to bring down the old European aristocracies. Furthermore, expansion into such vast new regions inspired many technological innovations and large-scale application of new techniques. In the same way, Mars would offer virgin territories and technical challenges that would lead humanity to a new vision of their relationship with nature and their attitude towards their own kind.

These may be interesting perspectives, but they should not be allowed to hide another major question, this time of an ethical nature, which greatly concerns scientists. In his excellent book *Pale Blue Dot*, published in 1995 just one year before his death, Carl Sagan clearly described what was involved here. Even if it turns out to be feasible from the technical point of view to terraform Mars or some other body, is this sufficient reason for actually doing so? Would it not be preferable to study and understand another world before changing it? And given the relatively short lifespan of our political and economic institutions, is it reasonable to undertake such a long-term project, when it might be interrupted at any moment, leaving the planet disfigured for all eternity? Should we not be conserving our Solar System in its present state for future generations, who might find better uses for it? And above all, in view of the terrible destruction we have already inflicted upon our own planet, is it wise to consider interfering with another world?

There is no way of answering these questions today, for they stand far outside our present preoccupations. But sooner or later, future generations will have to face them. In *Red Mars*, American science fiction writer Kim S. Robinson provides us with a wonderful illustration of this future debate. The book is the first in a trilogy of monumental proportions, together with *Green Mars* and *Blue Mars*. They follow through the various stages in colonising and terraforming Mars, undertaken by human beings in the twenty-second century. Robinson's trilogy is based on the work of the McKay group (presented in the previous section), and succeeds technically, as well as tackling social and economic questions which will confront our civilisation over the coming two centuries. All the problems involved in terraforming are brilliantly brought out in *Red Mars*, through the characters of geologist Ann Claybourne, and physicist Sax Russell. During a decisive meeting about the future of Mars, the confrontation between these two characters deserves to be quoted in any anthology. The geologist passionately opposes any attempt to change the planet, accusing the project's supporters of irresponsibility. In her view, they want to play God, without any respect for nature, and deliberately destroy the beauty of a landscape billions of years old, which could teach us so much about the past of our Solar System. The physicist's reply is equally profound:

> The beauty of Mars exists in the human mind; without the human presence it is just a collection of atoms, no different from any other speck of matter in the universe... the whole meaning of the universe, its beauty, is contained in the consciousness of intelligent life... we are the consciousness of the universe, and our job is to spread that around, to live everywhere we can. It is too dangerous to keep the consciousness of the universe on only one planet, it could be wiped out... we can transform Mars and build it like you build a cathedral, as a monument to humanity and to the universe. We can do it, so we will do it.

Naive optimism, dangerous illusion or prophetic vision? The answer may come in several centuries!

The frontiers of the Solar System

The main targets for space programmes in the coming century will clearly be near-Earth space, the Moon, Mars and its moons, and nearby asteroids. This selection is dictated as much by economic potential as by accessibility. The stakes include raw materials, such as water and volatile elements in the carbonaceous chondrites, and metals in the iron asteroids; high technology products manufactured in weightless, high-vacuum conditions; and energy from the Sun or thermonuclear fusion of lunar helium-3. Naturally, such projects will be subject to the vicissitudes of social and economic fluctuations prevailing on our planet. As regards opening up the Martian frontier by terraforming the planet, followed by massive colonisation, this will surely not happen before several centuries have elapsed. The same goes for exploitation of the many bodies lying in the main asteroid belt, between the orbits of Mars and Jupiter.

At the present time, other bodies in the Solar System are of interest only for pure research. Unmanned exploration of these objects will help us to understand the genesis and evolution of the Solar System.

Apart from Mars, the other planets are of little practical interest. High temperatures and pressures in the Venusian atmosphere make it impossible for people to survive there. Only terraforming could render it inhabitable, thereby providing a living area equivalent to the whole surface area of our globe. However, no such project seems feasible today. Unless some technological miracle is achieved, or an ingenious solution put forward, Earth's twin planet (by its mass) will always be inaccessible to mankind.

Mercury is the planet orbiting closest to the Sun and would appear to be slightly less hostile than Venus. It shares many features with the Moon, although it is just a little bigger: there is no atmosphere, its surface morphology is similar (dimpled with meteoritic craters), and there is a huge contrast between daytime and night-time temperatures, in fact, the most extreme variation of any object in the Solar System. The Sun's light is on average six times more intense at the distance of Mercury than it is in when it reaches our own orbit. When the Sun is at

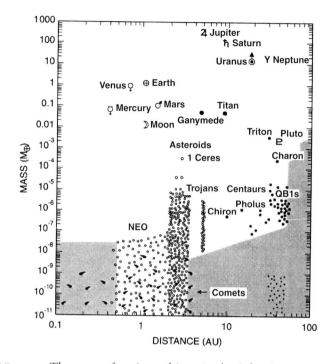

FIGURE 1.30   The mass of various objects in the Solar System as a function of their distance from the Sun. Masses are expressed in terms of the
mass of the Earth 1 M$_{\oplus}$ = 5×10²¹ tonnes, and distances in astronomical
units, where 1 AU = 150 million kilometres. Bodies indicated are the
nine planets, their most massive satellites (the Moon, Ganymede,
Titan, Triton and Charon), the most massive asteroid in the main asteroid belt (Ceres) and several representatives of various families of
smaller bodies in the Solar System, i.e., the NEOs (Near-Earth
Objects), the main asteroid belt, the Trojan asteroids, the comets and
their reservoir in the Kuiper belt. (Adapted from Rettig, T. & Hahn, J.
(1996) *Completing the Inventory of the Solar System.*)

the zenith, ground temperatures attain values around 450°C, whilst
the opposite hemisphere is condemned to darkness and a freezing
−170°C. In such conditions, the only potential interest in the planet
lies in the huge amounts of solar energy available in its neighbourhood (see Chapter 2).

Far beyond the orbit of Mars and the asteroid belt, at distances between 5 and 30 AU, are the giant planets of the Solar System: Jupiter, Saturn, Uranus and Neptune. They are much more massive than Earth, by a factor of 15 in the case of Uranus and Neptune, 95 for Saturn, and 317 for Jupiter, the most massive of all. They have a quite different chemical composition to the four innermost planets. Like the Sun, they are mainly composed of light gases, hydrogen and helium. It is impossible to see what lies hidden beneath their thick atmospheres, in which pressures rise steadily with depth. On 7 December 1995, a module launched by the American probe Galileo plunged down into Jupiter's atmosphere, transmitting observations for about an hour, before the tremendous pressure, 20 times the pressure on the terrestrial surface, made it inoperative. Several thousand kilometres down into the Jovian atmosphere, pressures reach values thousands of times greater still.

According to current models, the giant planets have a rock and ice nucleus about the size of Earth. It appears that the cores of Saturn and Jupiter may be surrounded by a metallic hydrogen mantle. This is a state of matter which can only exist at pressures several million times greater than the pressure on Earth. In these conditions, only the outer regions of these planets' atmospheres could ever present any practical interest. For they are rich in helium-3, an element which has been retained by their huge gravitational attraction. In an earlier section, we saw that fusion of lunar helium-3 would cover our energy needs on Earth over one or two millennia. In contrast, the accessible quantity of helium-3 around the giant planets (down to a depth where the atmospheric pressure becomes ten times the terrestrial pressure) would satisfy the needs of our civilisation for several billion years.

Unmanned probes could extract helium-3 from the giant planets. A small nuclear reactor would supply the energy needed to extract helium and separate the helium-3 and helium-4 isotopes. The precious helium-3 would be cooled and stored in liquid form. The much more abundant helium-4 would be heated to fill balloons deployed by the probe, allowing it to float in the atmosphere during the process. Once it had filled its reservoir with helium-3, the probe would begin the return journey to Earth, taking several long years. However, Jupiter's high mass would make it very difficult to escape from its tremendous gravitational grasp. Extraction of Jovian helium-3 would not be economically viable. Even if the probe were propelled by thermonuclear fusion, an exceptionally efficient energy supply, it would have to burn up a large part of its own helium-3 load. The other giant planets have smaller gravitational fields and their helium-3 contents may well be more easily accessible (in particular, those of Uranus and Neptune).

As a result of their weak gravity, the smaller bodies of the Solar System are of great practical interest. Three of the Jovian satellites harbour the largest fresh water reserves in the Solar System. Beneath their thick crusts of ice lie oceans of liquid water, 100 kilometres thick on Europa, and more than 500 kilometres thick on Ganymede and Callisto. The amount of water on Europa is comparable with all the oceans of our own planet, whilst Callisto and Ganymede contain 30 times as much.

It is hard to know whether mankind will one day turn to the Jovian satellites for water supplies. It is not obvious today how they could be extracted and transported across the Solar System. Moreover, much water is contained collectively in the form of ice in the myriad small bodies which populate the main asteroid belt. These also contain volatile elements and metals. The advantage in exploiting these resources lies in the small size of the asteroids, making it much easier to attain their wealth and transport the results back to the inner Solar System. Many science fiction stories relate the adventures of future astrominers, seeking their fortune in the spatial Far West of the third millennium.

Mining resources are not the only attraction of the asteroids. Some have suggested that mankind could colonise them in the distant future. Colonists would live either on the surface of the body, or on its inner walls, once it had been excavated. In the first case, they would protect themselves from the cold and cosmic rays by transparent plastic domes containing a suitable atmosphere (in terms of chemical composition, temperature and pressure). The asteroid's resources would allow them to recreate a totally self-sufficient biosphere. However, they would have to adapt to the weak gravity of their new world, which would be hundreds or thousands of times less than Earth's gravity. Those visiting the mother planet would find it extremely difficult to bear its crushing attraction. The same problem would not affect settlers who chose to inhabit the inner walls of the hollowed out body. Just as in O'Neill's cylinders, the asteroid's rotation would simulate a suitable gravity by centrifugal action (at least in equatorial regions).

Several thousand objects bigger than a few kilometres across orbit in the main asteroid belt. Taken together, their areas equal the whole land area on Earth. In his recent book *Mining the Sky*, American planetologist John Lewis estimates that the resources in this zone could satisfy the needs of a population millions of times the present Earth population. The Trojan and Greek asteroids, slightly further out, contain even greater resources. These objects are located at Lagrange points L4 and L5 of the Sun–Jupiter system. In other words, they are always at the same distance from Jupiter. The Trojans precede Jupiter by an angle of 60° in its orbit about the Sun, and the Greeks follow behind by the same angle. The total mass of these objects is three or four times greater than the mass of the asteroids making up the main belt.

Beyond the orbit of Neptune, at more than 35 AU from the Sun, extends a vast region mainly populated by comets. These balls of ice and dust, several kilometres in diameter, sometimes leave their icy domain to visit the inner regions of the Solar System. This happens when their orbits are perturbed by the gravitational field of a giant planet (or another star approaching the Sun). The Sun's heat tears gases and dust from their surface, forming a halo of light and a tail several million kilometres long. They then become the most spectacular objects in the night sky, rather like Hale–Bopp, which monopolised the attention of star-gazers over the first few months of 1997.

The idea that the disk of the Solar System might be extended out to several tens of astronomical units by a belt of comets was first put forward in the middle of the twentieth century by the Irish astronomer Kenneth Edgeworth and the Dutch scientist Gerard Kuiper. At the time, the only object known in this region was the little planet Pluto, discovered in 1930 by Clyde N. Tombaugh. Its satellite Charon, about half the size of the Moon, was discovered in 1978. Fifteen years later, in 1992, an object measuring about 200 kilometres across was discovered beyond the orbit of Pluto. Today, more than forty objects have been observed in this trans-Plutonian family, known as the Kuiper belt. It is estimated that there may be more than forty thousand objects of diameter greater than 1 kilometre orbiting in this region, making them a thousand times more numerous than their counterparts in the main asteroid belt. Taken together, the estimated ten billion comets in the Kuiper belt have a total mass comparable with the mass of Mars.

Finally, far beyond the Kuiper belt, a vast spherical cloud of comets encircles our Solar System. The so-called Oort cloud (named after the

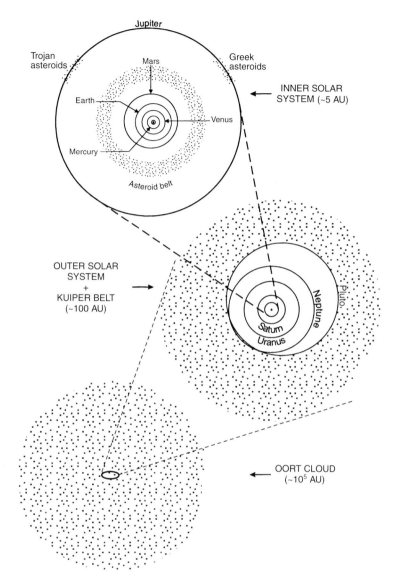

FIGURE 1.31   Respective sizes of the inner Solar System (within the asteroid belt, a few astronomical units from the Sun), the outer Solar System
(the Kuiper belt extends out about a hundred astronomical units), and
the Oort cloud (of radius almost 100 000 AU).

Dutch astronomer who first made the hypothesis) extends between 10000 and 70000 AU from the Sun. Resources in the Kuiper belt represent only one thousandth of those in the Oort cloud, which is estimated to contain about ten trillion comets! This is more than there are stars in our Galaxy! Despite their large numbers, the average distances separating these comets are comparable with the distance from the Sun to Earth. Their total mass is roughly that of one of the giant planets Uranus or Neptune.

The Kuiper belt and, in particular, the Oort cloud contain the largest reserves of volatile elements in the Solar System. These distant regions are totally inaccessible today but they may well be the new frontier in the fourth millennium. Will people live in these places? Who would wish to settle in permanent darkness, so far from the light and heat of our life-giving star? Seen from the Oort cloud, the Sun would be no brighter than the other stars in our night sky. The distance out to the edge of the Oort cloud is about a quarter of the way to our nearest stellar neighbour. However, some have already suggested that colonising the Oort cloud would be a natural step in our cosmic adventure, before moving on to the next objective: travel to the stars.

ROUTE TO THE STARS

Too low they build, who build beneath the stars.

Edward Young, *Night Thoughts*.

The Earth, that is sufficient,
I do not want the constellations any nearer
I know they are very well where they are
I know they suffice for those who belong to them.

Walt Whitman, *Leaves of Grass, Song of the Open Road*.

A rocket pulls away from its launch pad and rises majestically into the sky. This magnificent sight must surely be one of the most striking images to be retained from the twentieth century. Rockets have opened up the way to space, allowing humanity its first few steps in the vicinity of Earth, and then across interplanetary space. But the stars still lie beyond our grasp, as inaccessible as they ever were before the advent of space travel.

Will we one day cross the great ocean of interstellar space, and if so, how? How many decades, centuries or millennia will we need before we can reach at least the nearest stars? Today, we have no answer to these questions. The technical difficulties seem so daunting that they have not even been the subject of official studies. Interstellar travel still animates the dreams of some adolescents and science fiction writers, but contemporary science has preferred to postpone it indefinitely into the future.

In the 1950s and 1960s, this was not the case. Over a period of about twelve years, mankind had succeeded in controlling nuclear energy, building rockets and designing computers. Such an explosion of technical achievement could only encourage those with hopes for interstellar travel. Engineers and physicists turned to the new challenge with great enthusiasm, investigating some novel ways of casting a line out to the stars. Despite an admirable display of ingenuity, their efforts only served to evidence the difficulties involved. It is quite obvious today that, without some technological miracle, the route to the stars will be long... very long. In all probability, we shall not engage upon it for the next couple of centuries.

The interstellar ocean

We know that the Universe is immense. However, it is always hard to comprehend the real vastness of the distances separating us from the stars. Light travels the 400 000 kilometres separating the Moon and Earth in little more than one second. Neptune, the penultimate planet of the Solar System, gravitates ten thousand times further from Earth than the Moon, at 4.4 billion kilometres. Light covers this distance in about 4 hours. The nearest star, Proxima Centauri, is then ten thousand times further from us than Neptune. This time, light takes 4.3 years to cover the 40 000 000 000 000 kilometres (40 trillion kilometres) which lie between us and our stellar neighbour. At 30 000 light-years from Earth, almost ten thousand times further than Proxima Centauri, we come to the hub of the vast collection of a hundred billion stars which we call the Galaxy. Distances to nearby galaxies are measured in millions of light-years, whilst distant galaxies have been glimpsed billions of light-years away.

When Albert Einstein formulated his theory of relativity in 1905, it was realised that the speed of light in a vacuum represents an ultimate limit for motion of any material object. That is, no material thing can move faster than 300 000 kilometres per second (more precisely, 299 792 kilometres per second). This speed limit, denoted by c, posed serious difficulties for science fiction writers. What was to become of their galactic empires? We shall return to this point later in the chapter, but it is clear that our ambitions for interstellar travel must be restricted to the closer cosmic suburbs, at least for the time being.

Fortunately, the stage is not completely deserted. Twenty-six stars lie within a radius of 12 light-years from the Sun. Only three of these are bigger and brighter than our own star. The others are mainly small red stars. Our closest stellar neighbour, Proxima Centauri, is a tenth

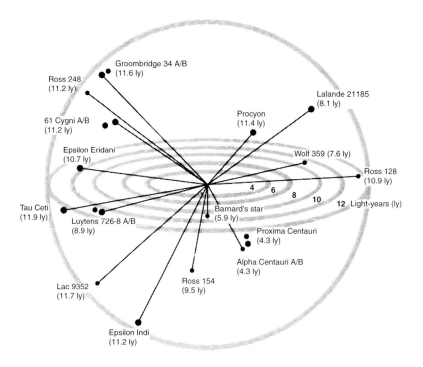

FIGURE 2.1   Positions and distances (in light-years) of the Sun's nearest stellar neighbours.

the mass of the Sun and a thousand times less bright. It belongs to a triple star system, rotating in a rather large orbit about Alpha Centauri (α Centauri) A and B, two stars very similar to the Sun, which form a close pair, 4.4 light-years from Earth. Just a little further, at 5.9, 7.6 and 8.1 light-years from the Sun, are Barnard's star, Wolf 359 and Lalande 21185, three red stars slightly more massive than Proxima Centauri. At 8.6 light-years, we encounter the Sirius double system. Sirius A is a white star, 2.3 times more massive and 23 times brighter than the Sun. Its intrinsic luminosity combined with its proximity make it the brightest star in the sky. Sirius B is almost as massive as the Sun but a thousand times less luminous and a hundred thousand times smaller. In fact it was the first *white dwarf* ever discovered. Finally, among the remaining eighteen stars, we find four pairs, including Procyon A and B, which are very similar to the Sirius system.

Unlike the planets of the Solar System, the stars in our vicinity do not form a group. They are so far from one another that they are not linked by gravity (except within double or multiple systems like Sirius or α Centauri). Since they do not have the same motions in the Galaxy, their mutual separations can vary considerably with time. This implies a gradual evolution of the solar neighbourhood over a time scale of tens or hundreds of thousands of years. Thus, Proxima Centauri has not always been our nearest neighbour. It became so about 33 000 years ago, replacing the double system Gliese 65. In 32 000 years, it will in turn be replaced by the small red star Ross 248, which is presently at 10.8 light-years from the Sun, but will come within 2.9 light-years. Four other stars will make an even closer approach over the next million years. Amongst these, DM+61366 will graze the Solar System at a distance of only 0.3 light-years (almost fifteen times closer than α Centauri). According to calculations, this should happen in the year 814 872.

Even in the best cases, distances to stellar neighbours are measured in light-years. This is fortunate, because a densely populated neighbourhood could have disastrous consequences for our Solar System. Indeed, as we have seen in the previous chapter, the close approach of a star could perturb the cloud of Oort comets which swarms around us, sending many projectiles towards the inner regions of the Solar System. If one of these collided with our planet, it might cause the extinction of many living species, just as we believe the dinosaurs may have been wiped out 65 million years ago (see Chapter 3).

Our cosmic isolation, so beneficial for the stability of the Solar System, constitutes the main obstacle to interstellar travel. In order to reach the nearest stars in a reasonable time, we would have to travel at a minimum of one tenth the speed of light (0.1c). Such high speeds are well beyond our present technological capabilities. Even extrapolating our knowledge to its very limit, it is hard to imagine that we might achieve these speeds in the foreseeable future.

Steeds for space travel

The only way known today for moving through empty space is based on the principle of action and reaction, first formulated by Isaac Newton in the seventeenth century. By simply projecting an object in

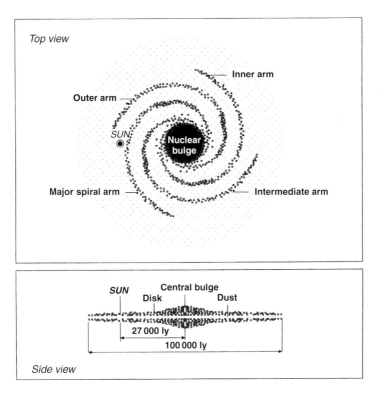

FIGURE 2.2   The position of the Sun in the Milky Way, a spiral galaxy about
100 000 light-years across. Top and side views are shown. The Sun is
located 27 000 light-years from the galactic centre. A layer of dust, settled
on the galactic plane, absorbs most of the light from that direction.

one direction, we feel a push in the opposite direction. The greater the
mass of the object and the speed with which it is projected, the faster
the motion imparted to us. This is the basic principle of the classical
rocket, which comprises the following elements: payload (including
the motor), propellant (the mass to be expelled during travel), and fuel
(burnt by the motor to accelerate the propellant to the desired speed).
Expelled after combustion, the fuel often serves as propellant too.

Classical rocket theory was founded by Konstantin Tsiolkovsky, the
father of astronautics. The celebrated rocket equation, almost as
famous as $E = mc^2$, can be attributed to him. According to this equation, the terminal velocity V of a rocket equals the speed v of ejected

propellant multiplied by the natural logarithm of the ratio of initial to final masses M/m: $V = v \ln (M/m)$. In order to reach a high terminal velocity, the mass of propellant $M - m$ and its ejection speed v must be as large as possible. The final mass m can also be reduced, but below a certain limit the mission no longer serves any purpose because it could not carry any equipment. Of course, the limit is much higher for manned flights than for unmanned flights.

Ejection speeds depend on the method of propulsion. In the rest of this chapter, we shall present several systems of propulsion already operational, or envisaged for the near future. All these systems use energy sources, such as chemical, nuclear or some other form of energy, which convert part of the fuel mass into energy. The larger the converted fraction (known as the efficiency of the process), the higher the ejection speed. Chemical energy sources have low efficiency (below 10^{-10}) and produce ejection speeds of a few kilometres per second. Nuclear energy sources are millions of times more efficient, reaching about 7×10^{-4} for fission reactions with uranium isotopes and 5×10^{-3} for fusion reactions with light isotopes of hydrogen. Theoretically, these values lead to ejection speeds of order $0.01c$ to $0.1c$ (typically, a few hundredths of the speed of light). In practice, this theoretical performance is limited by the way in which combustion energy is transferred to the propellant.

One way of increasing the terminal velocity of the rocket is to play around with the other term in Tsiolkovsky's equation, namely the ratio of initial to final mass. Unfortunately the natural logarithm is a slowly varying function. The mass ratio must be greatly increased to obtain a modest improvement in the speed. This is made clearer by writing Tsiolkovsky's equation in the form: $M/m = \exp (V/v)$. The mass ratio increases exponentially with terminal velocity V (for fixed ejection speed). Hence, a mass ratio of 10 is needed to give $V = 2.3v$, and 100 to double that speed, i.e., to give $V = 4.6v$. A ratio of 1000 merely triples the first value, giving $V = 6.9v$, and we see that it soon becomes enormous, requiring huge quantities of fuel (Fig. 2.3). The reason for this exponential increase is that the classical rocket must transport its own fuel and propellant. The more it carries, the more it needs in order to do the carrying. Generally speaking, the final mass is only a tiny fraction of the initial mass, the latter being totally dominated by the mass of fuel and propellant.

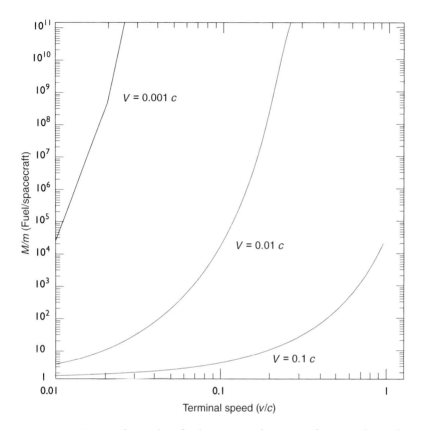

FIGURE 2.3   Ratio of initial to final mass as a function of terminal speed
for a classical rocket. The initial mass is primarily due to the fuel, whilst
the final mass is just that of the spacecraft. Speeds are given as a fraction
of the speed of light c. To reach nearby stars within reasonable times,
the terminal speed must be at least $0.1c$. The three curves correspond to
propellant ejection speeds of $0.001c$, $0.01c$ and $0.1c$. Only in the third
case are useful terminal velocities attained with a reasonable mass ratio.

The idea behind rockets like Saturn V, Ariane or Energia can be
traced back to fireworks used by the twelfth-century Chinese. These
rockets draw energy from chemical reactions which suddenly re-
arrange the electrons around atomic nuclei. Such chemical reactions
take place between two substances, the fuel and an oxidizer, in solid or
liquid state. In the first case (the direct descendent of the fireworks),

fuel and oxidizer are stored in the form of a fine powder. Liquid fuel rockets are generally more complex. The fuel must now be injected into the combustion chamber as a jet of fine droplets, at high temperatures and pressures. In both cases, the fuel is rapidly heated to a temperature of several thousand degrees, causing it to vaporise and dilate. It leaves rapidly by way of the exhaust nozzle, the only exit, so that the rocket receives a thrust in the opposite direction.

Ejection speeds obtained in this way depend on the gas temperature and the 'weight' of its molecules. The higher the temperature and the lighter the molecules, the greater will be their speed. These two factors are in turn determined by the type of fuel. Using today's solid fuels, speeds of 3 km/s can be attained. The most powerful propellant at present used in the space programme (and in the main motors of the space shuttle) is a mixture of liquid hydrogen and oxygen, which react to produce water vapour. The ejection speed is 4.5 km/s, close to the maximum performance of 5 km/s predicted by theory for this propellant.

Today, theoretical performances are known for all chemical fuels. Some chemical combinations can give rise to slightly higher ejection speeds than the hydrogen–oxygen mixture, reaching 7 km/s. This is the case for the liquid fluorine–LiH_2 mixture and also the oxygen–BeH_2 mixture. These combinations are highly explosive and could doubtless never be used to propel a rocket. Chemical fuels that perform better still have been suggested. Their greater efficiency remains to be proved, but whatever happens, ejection speeds obtained in this way will never exceed 20 km/s.

It is clear that chemical rockets will never take us to the stars. To reach a terminal speed of 10 000 km/s (3% the speed of light), with optimal ejection speed of 10 km/s, we would need a mass ratio of exp 1000. This means that for a payload of just one atom, we would have to eject 10^{434} fuel atoms. However, the observable Universe only contains 10^{80} atoms. Fortunately, there do exist more efficient energy sources.

A technique for which the technology has already been developed and which can attain high ejection speeds is ion propulsion. An electrical generator supplies the energy required to ionise atoms in the propellant, tearing electrons away from their nuclei. The resulting ions have positive charge and can be accelerated to high speeds by a powerful electric field. The latter is also supplied by the generator on board. Obviously, if the electrons were allowed to accumulate in the rocket, it would rapidly become negatively charged. It would then attract the ejected beam of positive ions, thereby reducing overall performance. For this reason, electrons are also accelerated towards the output nozzle, where they meet up with outgoing ions to form an electrically neutral beam.

Ion propulsion motors have already been built and tested in near-Earth space. Ejection speeds currently vary between 25 and 100 km/s, whereas the best performance which could theoretically be obtained by this technique is estimated at about 1000 km/s, fifty times greater than the best achievements of chemical propulsion.

Since 1940, nuclear fission has been used to produce energy, and is today the source of energy in our nuclear power stations. An ambitious programme aiming to build fission rocket motors was carried out in the United States in the 1960s. Experimental prototypes (NERVA, Kiwi and Rover) were built and tested in the laboratory. However, none of these ever took to the air. The programme, which cost several billion dollars, was dropped in the 1970s.

Energy released by the fission reactor is used to heat the propellant (generally hydrogen) to a temperature of several thousand degrees. The higher the temperature, the faster the propellant will be expelled through the nozzle. The state of the reactor core determines this temperature. Gas core reactors yield higher temperatures than solid or liquid core reactors. Ejection speeds vary between 5 and 11 km/s for a solid core reactor and can reach 30 to 70 km/s for a gas core reactor.

Despite its great efficiency as an energy supply, nuclear fission reactors alone could not produce the high ejection speeds needed to engage upon interstellar travel. Performance is limited by propellant temperature, which cannot exceed certain bounds (otherwise the reactor core may melt down or vaporise). However, in combination with ion propulsion, the situation is more promising, for there are no temperature restrictions on the ion propulsion motor. (Once accelerated, ions are channelled along by the electric field without touching the nozzle walls.) If the ion propulsion motor were supplied by a fission reactor, we could reach the nearest stars. However, exorbitant amounts of fissionable material (e.g., uranium) would be needed for the journey. A mass ratio of 5 million would be needed to reach Proxima Centauri in one century with speed 0.05c and ejection speed 1000 km/s, and this only for the outward journey. A payload of 10 tonnes implies 50 million tonnes of uranium, thousands of times greater than today's world production of this element.

Nuclear fusion is a more efficient way of converting energy than nuclear fission. Although explosive fusion is feasible today, controlled fusion has not yet been achieved. As observed in the last chapter, experience over the previous forty years suggests that it will not be accomplished for several decades to come. In a rocket propelled by controlled fusion, hot plasma escapes from one side of the reactor, either directly or after further acceleration by a magnetic field. Fairly high ejection speeds could be obtained in principle, of the order of several thousand kilometres per second. Such a performance surpasses the best theoretical achievement of ion-propelled rockets (supplied by fission reactor). Fusion reactors may well be much heavier than fission reactors. In fact, it is hard to imagine anything less than a hundred tonnes. But starships with almost reasonable initial masses could then be conceived, due to their superior performance.

Whilst controlled fusion remains in the research stage, physicists have been investigating another idea: using energy from nuclear explosions.

The Orion project and Dyson rockets

The idea of using energy from explosions to propel a vehicle is not new. After all, car engines are based on an explosive chemical reaction, albeit confined within the engine (and the cause of all its complexity). At the end of the last century, the German engineer Hermann Ganswindt designed vehicles pushed along by chemical explosions. Half a century later, the first atomic bombs demonstrated the extraordinary efficiency of nuclear explosives. But how could anything survive the cataclysm produced by such explosions? How could the tremendous energies they release be channelled and controlled?

Physicists involved in the Manhattan project, which designed the first atomic bombs, began to think about these problems at the end of the 1940s. In 1947, the mathematician Stanislaw Ulam and his colleague Frederic Reines put forward the idea of *pulsed nuclear propulsion* in a report prepared at the Los Alamos laboratory in New

Mexico. The basic element is a metal plate, called the pusher plate, covered on one side by a graphite layer which can absorb large quantities of heat (i.e., it has high specific heat capacity, in the jargon of physics). For a fraction of a second, this plate could resist the heat released by a nuclear explosion occurring a few dozen metres away. Some of the vaporised debris from the bomb, projected at several tens of thousands of kilometres per second, would bounce off the plate, giving it a violent push in the process. If the pusher plate were connected to a spacecraft by some mechanism able to damp such powerful impacts, its motion would be transmitted to the starship. The latter could then be propelled by explosions repeated at regular intervals.

Experiments carried out at the beginning of the 1950s, in the context of the American military nuclear programme, proved that the suggestion was not completely absurd. Ordinary materials such as steel and aluminium could resist temperatures of several hundred thousand degrees for a few milliseconds without serious ablation at the surface. In 1955, encouraged by these results, Ulam and Cornelius Everett made the first theoretical calculations for pulsed nuclear propulsion. They considered a 12 tonne vehicle, equipped with a reflecting plate 10 metres across and a hundred or so weak nuclear charges. One bomb was to explode every second, 50 metres from the plate. This would impart a final speed of 20 km/s to the rocket.

These preliminary calculations served as basis for the Orion project undertaken in 1958. It was begun just after Sputnik had been launched and before the American programme for chemically propelled rockets. Orion cost around 11 million dollars, with a team of about forty physicists working over a period of seven years. The group leader was Theodore Taylor, who had been previously engaged in nuclear weapon design at Los Alamos.

A large part of the work carried out by the Orion team concerned the design of suitable shock absorbers, able to cushion the explosive impacts. Ablation tests were also made for various materials. These probably used high temperature plasma generators rather than real nuclear charges. Although the details remain largely classified, it seems that Taylor's team solved the main technical problems in a satisfactory way. A single experimental vehicle, called *Put-put*, was built and tested in California. It reached a height of 60 metres with the help of chemical explosives.

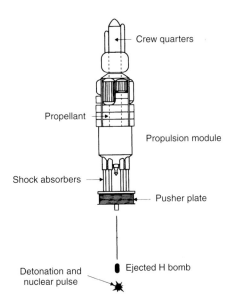

FIGURE 2.4   The Orion project interplanetary vehicle.

The Orion team designed several interplanetary rockets. A typical rocket would have a mass of a few thousand tonnes, and would use several thousand low-intensity nuclear charges, of between 0.01 and 10 kilotonnes. (The Hiroshima bomb was equivalent to 20 kilotonnes.) Bombs were to explode at regular intervals of a few seconds and could accelerate the vehicle at close to 1 g (terrestrial acceleration due to gravity at sea level). With final speed between 20 and 100 kilometres per second, the vehicle could do a return trip to Mars in less than a year.

The Orion project never got the green light from the American authorities. The 1963 treaty banning nuclear tests in the atmosphere or in space was a fatal blow. Chemically propelled rockets had finally won the day. In 1968, the physicist Freeman Dyson, one of the key members of the Orion team, wrote: 'Our plan was to send ships to Mars and Venus by 1968, at a cost that would have been only a fraction of what is now spent on the Apollo program. . . . I believe that fundamentally a Saturn V bears the same relation to an Orion ship as the majestic airships of the 1930s bore to the Boeing 707.'

Orion focussed on *interplanetary* travel by means of nuclear fission explosions. Despite his frustration, Dyson wasted no time in extrapolating the idea to interstellar travel. In 1968, he suggested that thermonuclear fusion explosions could propel a manned vehicle to nearby stars. His preferred project involved a fairly compact space vehicle with mass of about 20000 tonnes, connected to a pusher plate of similar mass by shock absorbers 100 metres long.

The starship was to carry three hundred thousand thermonuclear bombs, each weighing one tonne and with an explosive power of one megatonne (fifty times the Hiroshima bomb). With an explosion every three seconds, the vehicle would reach a terminal speed of 10000 km/s, or 3% the speed of light, after ten days. Cruising through space at these speeds, it would take about a century to reach the closest stellar system, α Centauri. On arrival, it would have to decelerate. For this purpose, a second stage would have to be added, bringing the total mass to one and a half billion tonnes.

More nuclear bombs were needed just for this outward journey than the whole world nuclear arsenal of the day. Dyson's proposal was of some academic interest, intending to demonstrate that the nearest stars could be reached within a reasonable period of time using only the technology available in the 1960s. In any case, there can be no doubt that the project offered an excellent way of eliminating the world's nuclear arsenal!

It is interesting to estimate the cost of such a mission. Dyson evaluated it at a few hundred billion dollars, comparable with America's gross national product at the time. Bearing in mind that, in peace time, no project could be undertaken that would absorb more than a few per cent of a nation's resources, Dyson suggested that his project would not become economically viable for another two centuries. He based this estimate on an annual economic growth rate of 4%, typical of the auspicious 1960s. Such a figure would have to be significantly reduced today. Generally speaking, the same considerations apply to all the interstellar travel projects we shall examine in the present chapter, which do not depend on technology that is too futuristic. It shows that, unless some major technological revolution occurs, it will be difficult to envisage interstellar travel within the next couple of centuries.

Daedalus: the project

Research into controlled nuclear fusion at the beginning of the 1970s completely transformed ideas of pulsed propulsion for interstellar travel. Instead of megatonne charges, requiring extremely heavy-duty structures to absorb the shocks they produce, micro-explosions were considered, thousands of times weaker than the Hiroshima bomb. These tiny explosions were to be triggered by a technique called *inertial confinement*. A miniscule pellet of light isotopes is bombarded by powerful photon or electron beams. The idea is that beams converge on their target from several directions at once, compressing and heating it to several hundred million degrees so that fusion can occur. The method is under study at the moment in laboratories around the world. Although no convincing results have been obtained for the time being, it remains a promising approach to thermonuclear fusion.

Unlike the massive explosions in Dyson's rockets, micro-explosions induced by inertial confinement would occur inside the starship rather than outside. The pusher plate in the Orion project would be replaced by a magnetic field which would channel hot plasma from the explosion towards the exhaust nozzle. Part of the energy from the explosion would be recovered and used to supply the inertial confinement beams.

The Daedalus project of the British Interplanetary Society is based on these ideas. Founded in 1933, this society is one of the oldest space organisations in the world. Even in 1939 they published a study which went well beyond the capabilities of the day: to send a manned spaceship to the Moon. The spaceship was designed in the context of 1930s technology and bore little resemblance to rockets used in the Apollo programme. However, it proved that such a mission was technically feasible. The British Interplanetary Society launched the Daedalus project in the same spirit in 1973. It was named after Daedalus, the brilliant engineer of Greek mythology, father of Icarus, who made wings so that they could both escape from King Minos' prison, the Labyrinth. Alan Bond, of the Culham laboratory in Great Britain, led a team of thirteen physicists and engineers on this project over a period of five years. At present their work constitutes the only detailed study of an interstellar rocket.

Given the great difficulties involved in interstellar travel, the Daedalus team fixed itself the simplest possible objective: an unmanned flight which would fly by a nearby star without deceleration, in order to transmit observational data to Earth. A speed of about $0.1c$, 10% the speed of light, would guarantee a reasonable time scale for the mission, of the order of half a century.

Barnard's star was chosen as target. After α Centauri, this is the closest star to the Sun. It can be seen only with the help of a telescope, being a small ninth magnitude star in the Ophiucus constellation. It is in fact a red dwarf, a typical member of the class containing the least massive and most common stars in the Galaxy. It is only one tenth the mass of the Sun and two thousand times less luminous, with surface temperature never exceeding 3000 K. If it were placed at the same distance as the Sun, this star would appear two hundred times brighter than the full Moon and six times smaller. The main reason for selecting Barnard's star as the target for Daedalus was just that in the 1960s it was suspected of having a planetary system. However, as the project was coming to its end, the observations in question were shown to be inaccurate, and no further evidence has been gathered since then.

Daedalus would be propelled by energy from the fusion of the light isotopes deuterium (D) and helium-3 (^{3}He). In the previous chapter, we saw that this reaction has a specific advantage when planning Earth-based fusion reactors: the small amount of neutrons released ensures that the reactor walls will not become radioactive. As Daedalus would be unmanned, the problem here is not to protect life. However, energy released by uncontrolled particles might cause the motor to overheat. On Earth, nuclear power stations are easily cooled by means of large quantities of water, but this would be impossible in space. The mission designers therefore chose the deuterium–helium-3 reaction, although well aware that there is no helium-3 on Earth, nor on any of the telluric planets Mercury, Mars and Venus. These planets have weak gravitational fields, unable to retain such light elements as hydrogen and helium, which escaped into space during planetary formation. Moreover, the Daedalus team was not aware of results from lunar rock analysis, showing that helium-3 could be found on our satellite.

Accelerating Daedalus to one tenth the speed of light would require about 30 000 tonnes of helium-3 and almost as much deuterium. This amount of fuel could be extracted from the atmosphere of the giant planets, which have retained their light elements. The Daedalus team suggested the Jovian atmosphere as helium-3 supply, almost as

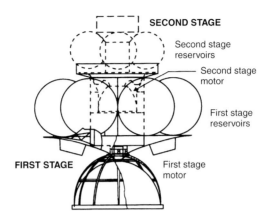

FIGURE 2.5   The Daedalus starship

ambitious a project as the interstellar voyage itself! A fleet of about a hundred extraction units would have to be sent out to Jupiter, equipped with mini fission reactors and suspended from giant balloons. They would have to operate for about twenty years in order to extract the necessary amounts, separating helium-3 from its heavier and more abundant isotope, helium-4. Other automated craft would operate as a shuttle between the balloons to collect the precious fuel. The best approach would be to refuel the starship near Jupiter, to avoid transporting thousands of tonnes of helium-3 across the Solar System. The Daedalus team selected the Jovian satellite Callisto, which is close enough to Jupiter without being inside its dangerous magnetosphere.

The starship would be 190 metres long and consist of two stages (see Fig. 2.5). The first-stage motor would be an enormous hemispherical chamber, 100 metres in diameter, as big as the dome in Saint Peter's basilica in Rome. Its inner wall would be covered with a battery of seventy-five electron cannons pointing towards the centre of the hemisphere. Two hundred fuel pellets (containing D and ^{3}He) would be injected into the motor every second. Here they would be instantaneously struck by the thunder of the electron cannons. One thousand terawatts would be discharged onto each pellet for just twenty billionths of a second. Compressed by this colossal power, the pellet would detonate immediately. Explosive fusion of deuterium and helium-3 would release an energy equivalent to several tonnes of TNT. Part of this energy would be recovered in order to supply the next beam. Frequent repetition of these mini explosions would liberate energy equivalent to about one megatonne of TNT every hour inside the motor.

Just above the main motor, surrounding the structural frame of the vehicle like a bunch of grapes, six spherical tanks measuring 60 metres across would carry 46 000 tonnes of fuel for the motor in the first stage. The fuel would be in the form of about twelve billion 2 gram pellets of deuterium and helium-3. In order to maintain these in a solid state, the pellets would be suspended in a liquid helium-4 bath cooled down to just 3 K (three degrees above absolute zero). When emptied, each tank would be ejected to lighten the load carried by the starship.

The second stage would be a reduced version of the first. Its tanks would contain 4000 tonnes of extra fuel. The payload would be located at the top of the rocket, in the form of a cylindrical module weighing 450 tonnes. This would include the on-board computer and instrumentation for the various experiments: eighteen probes designed to explore the interstellar medium and the Barnard system, astronomical equipment (including two 5 m telescopes), and mobile robots controlled from the starship computer, capable of carrying out repairs.

Daedalus: the mission

Daedalus would be assembled near an orbiting space station in the inner Solar System. Once completed, the starship would be placed in orbit around Callisto to fill its tanks with 50 000 tonnes of propellant extracted from the Jovian atmosphere. On the great day of the launch, a long stream of hot plasma would spurt out at 10 000 km/s. The motor would run for two years, imparting a speed of $0.07c$ (7% the speed of light) to the starship. The first stage would then be ejected and the second-stage motor would take over for another twenty months. Almost four years after its departure, the spaceship would reach its cruise speed of $0.12c$.

At this point, Daedalus would already be streaking away at 0.2 light-years (13 000 AU) from Earth. A radio signal would take five months to do the return trip from starship to ground control. It is quite clear that any interstellar mission, whether manned or unmanned, will have to be completely autonomous. The mother planet could not guide its progeny with such a time lapse. The mission would need an on-board computer at least as powerful as HAL in *2001: A Space Odyssey*. Right through this never-ending journey, the computer must control everything that happens on the starship. This includes instructing robots to carry out maintenance and reacting to unexpected difficulties. It must also analyse observational data collected by telescopes, particularly in the approach to Barnard's star. Using this data and instructions sent out from Earth (before or during travel), the computer must devise the best plan for effective use of the equipment available to it, in order to obtain as much information as possible. It must then launch the probes along appropriate trajectories, make a preliminary analysis of their observations and communicate everything to Earth. Needless to say, no computer exists today with these capabilities, enabling it to replace a human crew. Nevertheless, progress in the development of artificial intelligence could make this a realistic objective in the not too distant future.

Twenty-five years after its departure, telescopes on the starship would begin to scrutinise Barnard's star in the hope of identifying its hypothesised planetary system. Using 1970s technology, a giant planet like Jupiter would be visible a few years before the encounter, although a small planet like Earth would only be detected a few months before. In this respect, the Daedalus designers were not sufficiently optimistic regarding the extraordinary progress that would be made in astronomical observation. Indeed, giant planets have already been detected in the vicinity of nearby stars, and there is every reason to believe that telluric planets will also be detectable (if there are any!) well before the first interstellar vehicle has even been launched.

Launched upon its daring mission at 36 000 km/s, Daedalus would cross the Barnard system in the space of a few days. Probes sent out by the computer would have roughly the same speed, possessing little propellant with which to decelerate. Any encounter with a planet would last at best for a few minutes. Hence the need for very meticulous planning, so that as much information could be gathered as possible. This implies sending the probes out along suitable trajectories several years before the encounter. A last-minute launch would be too risky, and too costly in terms of fuel.

In this way, following a fifty-year voyage across interstellar space, everything would be ready for mankind's first encounter with the stars, through the intermediary of probes. However, one final threat must be faced: the presence of comets and asteroids in the Barnard system. As we have seen, these objects are extremely common in our own Solar System, and may well be a feature of the Barnard neighbourhood. Collision with even the smallest piece of interplanetary debris would be disastrous for the starship or its probes hurtling along at 36 000 km/s. The designers of the Daedalus project proposed the following method for avoiding such a catastrophe. Upon entry into the Barnard system, the main spaceship and each of its probes would throw out an enormous smoke screen which would precede them over a distance of several tens of kilometres. Particles in these clouds would have the same speed as the spaceship and would destroy any comet-sized object lying in their path before it could reach the projectiles at their heel. Of course, the smoke screen would have to be dispersed by means of explosions at the moment of flyby, so that observing equipment on board probes could get a clear view of the planets.

Such a brief encounter would not permit detailed analysis of the Barnard system. For example, only one hemisphere of each planet (the one facing the star) could be photographed during the few minutes of observation, since the other would be plunged in darkness. This restriction would not prevent detection of terrestrial-type life forms on one of the planets. In fact, oxygen could be detected by spectroscopy, and this would constitute a clear indication of living organisms of the kind that inhabit our own planet.

After several days, Daedalus would leave the Barnard system and pursue its course across interstellar space with no further objective. Its computer would have transmitted all the observational and experimental data back to Earth. It would be six years before the electromagnetic waves carrying this information could be sensed on Earth. So fifty-six years after the launch, an extra-solar system would have been 'visited' by Earthlings.

It is very unlikely that our first interstellar emissary will look anything like Daedalus. More promising ideas exist today, as we shall soon see. However, this project had the merit of showing in a quantitative way that deep space travel (albeit unmanned) really is possible with the kind of technology we will have at our disposal in the coming century: advanced artificial intelligence, thermonuclear fusion via inertial confinement and the knowhow involved in extracting helium-3 from the atmosphere of Jupiter. The only one of these three prerequisites which seems to go beyond our present capabilities is the provision of helium-3. It presupposes a great deal more experience in interplanetary travel and construction than has been acquired at the present time. On the other hand, there is no fundamental reason why we could not overcome this obstacle.

Antimatter. The most efficient fuel…

The Orion and Daedalus rockets presented in the preceding sections give a good illustration of what can be achieved in interstellar travel using thermonuclear fusion. The process converts matter into energy with about 0.5% efficiency, ejecting propellant at speeds v of order $0.03c$ (a few hundredths of the speed of light). In order to reach nearby stars within a reasonable time scale, flight speeds V of order $0.1c$ are required. The rocket equation shows that this is possible, provided that the mass M of propellant can be raised to ten or a hundred times the mass m of the spaceship. (In the case of Orion, $m = 20\,000$ tonnes, $M = 400\,000$ tonnes, giving a mass ratio $M/m = 20$. In the case of Daedalus, $m = 450$ tonnes, $M = 50\,000$ tonnes, giving a higher ratio $M/m = 110$ and a higher speed $V = 36\,000$ km/s.)

The most efficient process for converting matter into energy is the mutual annihilation of matter and antimatter. The concept of antimatter, invented in the 1930s by the British physicist Paul Dirac, has always retained mythical status in the eyes of the general public. In reality, antiparticles have the same mass as their ordinary matter counterparts and obey the same physical laws. However, their electrical charges and other physical properties are opposite. Hence the positron, the first antimatter particle to be detected, has a positive charge equal in absolute value to the charge on its counterpart, the electron, but with opposite sign.

When a particle meets its antiparticle, they annihilate. More exactly, they are transformed into photons. All their mass becomes energy, with 100% efficiency. Einstein's celebrated formula $E = mc^2$ shows that annihilation of 1 gram of antimatter provides as much energy as the fission of 5 kilograms of plutonium, equivalent to the twenty kilotonne Hiroshima bomb. Matter–antimatter annihilation is thousands of times more efficient than fission and hundreds of times more efficient than thermonuclear fusion.

The maximal efficiency of antimatter as an energy source has not escaped the notice of rocket designers. The first person to study this idea was German engineer Eügen Sänger at the beginning of the 1950s. The only antiparticle known at the time was the positron, obtained by decay of certain radioactive elements. Electron–positron annihilation gives rise to two gamma (γ) photons with energies several tens of thousands of times higher than the energy of visible photons. Sänger hoped that these photons could propel his photon rocket at almost the speed of light. Unfortunately, photons produced by annihilation are emitted in random directions with equal probabilities. Being electrically neutral, they cannot be focussed through an output nozzle by magnetic fields. Furthermore, they cannot be reflected by mirrors in the way that visible light can, because they are so small (i.e., have such short wavelengths) that they slip through the atoms of any solid material. Sänger spent the rest of his life seeking a way of channelling the γ photon flux produced by electron–positron annihilation, but in vain. Rather than a rocket, it was something more like a photon bomb which he obtained, and today, the idea is no longer given serious consideration.

In 1955, Emilio Segrè and his collaborators at Berkeley, California, discovered a second antiparticle, the antiproton. It annihilates in a complex way upon contact with a proton. Not only are γ photons produced, but also unstable particles called pions, some of which are electrically charged. This feature of proton–antiproton annihilation makes it relevant in the search for a system of propulsion. Sänger's photon rocket will never see the light of day, whereas a pion rocket remains quite feasible, at least in theory.

When a proton and antiproton annihilate one another, five particles are created on average: three electrically charged pions (π^+ and π^-) and two neutral pions (π^0). The latter decay instantaneously, in 10^{-16} seconds, yielding two γ photons. Such electrically neutral particles cannot be controlled by magnetic fields and are therefore useless for propulsion purposes. Charged pions are also unstable, but have longer lifetimes, viz., 2.8×10^{-8} seconds. They decay into neutrinos and charged muons μ^+ and μ^- (depending on the charge of the pion). The muons are in their turn unstable and decay into electrons, positrons and neutrons. Finally, positrons annihilate with electrons to yield γ photons.

Hence, at the end of the process, proton–antiproton annihilation transforms the two particles into γ photons and neutrinos. The latter are indeed ghostly particles. They interact little with matter and would certainly be of no use in propelling a rocket, in contrast to the charged intermediate particles. Pions and muons spurt out at high speeds, around $0.9c$, carrying almost half the available energy with them. Although short-lived, they could be used to propel an antimatter rocket.

It might be thought that the kinetic energy of these charged particles could be used directly, channelling them towards the exhaust nozzle by means of a magnetic field. This would produce extremely high ejection speeds, around $0.9c$. However, the mass of propellant ejected would necessarily be very low, in fact less than the mass of antimatter used in the first place. As we shall see in the next section, there is little hope of ever having more than a minute amount of antimatter. This means that only tiny rockets could ever be propelled in this way.

It turns out that the most effective way of converting the fabulous annihilation energy into rocket thrust is in fact rather conventional: it consists in using it to heat a fluid propellant and ejecting that from the back of the rocket. Several methods have been devised for transferring energy from the charged pions and muons to the propellant fluid. These techniques are rather complicated and some technical problems have not yet been solved, even on paper. Indeed, numerical simulations show that only a few per cent of the annihilation energy could be used for propulsion. In the best of all worlds, it is clear that the propellant could not be ejected at speeds greater than $0.5c$, half the speed of light. Faced with the hard reality of these technical constraints, Sänger's dream of a photon rocket remains a long way off indeed.

We should not be too disappointed though. The extraordinary efficiency of antimatter as an energy supply still promises some benefits. In fact, mathematical analysis of antimatter propulsion reveals that, in any kind of mission, there is an optimal ratio of propellant mass M to spaceship mass m. This optimal value is around five. Moreover, the same analysis shows that the quantity m_a of antimatter needed to fuel a mission at speed V is equal to a mere fraction of the spaceship mass: $m_a = m (V/c)^2$.

The results of this analysis are better understood if we apply it to some specific cases. Case one: a payload of one tonne is to be sent to α Centauri in less than fifty years. Our analysis shows that to achieve the required speed of 0.1c, we would need only four tonnes of propellant and about a dozen kilograms of antimatter! Second case: a Daedalus-type mission. As we saw in the last section, using D + ^{3}He fusion the spaceship would need 50000 tonnes of fuel to propel a payload of just 450 tonnes up to a speed of 0.12c. The same performance could now be achieved with only 2200 tonnes of propellant heated by annihilating 8.5 tonnes of antimatter. In other words, the amount of propellant is cut by a factor of twenty-five. The figures imply that interstellar missions could be accomplished by relatively small rockets. Indeed, the Saturn rocket and the space shuttle require precisely 2000 tonnes of propellant!

... and the most expensive!

Unfortunately, we are currently unable to produce anything like a few kilograms of antimatter, and we are unlikely to be able to do so for the next century. It is true that Dirac's theory suggests a perfect symmetry between matter and antimatter, implying equal abundances of each across the Universe. But at the present time, no trace of antimatter has been observed in the Solar System, the Galaxy, or even in any distant galaxy.

The mystery of the missing antimatter has still not been cleared up. At the end of the 1960s, Soviet physicist and dissident Andrei Sakharov suggested that the asymmetry between matter and antimatter might have originated in the hot, early Universe of the Big Bang. In his view, physical phenomena taking place at 10^{27} K might have slightly favoured matter creation. During subsequent annihilation, some matter would have remained when all antimatter had been used up. The tiny excess of matter surviving annihilation, about one particle in a billion, would have formed the Universe we observe today. The suggestion is based upon certain theories of modern microphysics which have not yet been corroborated experimentally. The dominance of matter over antimatter in our Universe has thus not yet found a confirmed explanation. In any case, even if there are regions of antimatter elsewhere in the Universe, they certainly lie beyond the stars of our Galaxy and are therefore unattainable.

Today, antimatter is produced in the laboratory by the reverse process to annihilation, in which energy is converted into matter. Rewriting Einstein's formula as $m = E/c^2$, we see that an enormous quantity of energy must be expended in order to get a small amount of antimatter and matter (in equal fractions, of course). In the big particle accelerators like those in CERN (Centre Européen pour la Recherche Nucléaire) in Geneva, or Fermilab in the United States, protons are accelerated to speeds close to the speed of light and then made to collide with a metal target. During collision, the kinetic energy of the protons is partially converted into matter in the form of particles, including a few antiprotons.

Antimatter production is still a highly inefficient process today. Less than one thousandth of the proton kinetic energy becomes antiprotons. Even worse, we are unable to capture and store more than one in every thousand antiprotons actually produced. With such low yields, it is no surprise that the cost of producing antiprotons should be so very high, for ridiculously small amounts of the final product.

The Fermilab accelerator in the United States can produce about 50 billion antiprotons per hour, or 3×10^{14} antiprotons per year. This annual production corresponds to one billionth of a gram of antimatter, or one nanogram. Fermilab running costs (including paying off the construction cost) are around 50 million dollars a year. Of course, none of the world's accelerators are devoted to antiproton production today, and efficiency could be increased by a factor of a hundred, for a tenth of the cost. Even in this optimistic view, annual production would only reach about one millionth of a gram (one microgram) for a cost of around 10 million dollars. This must surely be the most expensive thing that will ever be dreamt up! Just to produce one milligram (a thousandth of a gram) of antimatter today would vastly exceed the economic and energy potential of the whole planet.

Even if we were able to increase the efficiency of antimatter production, the quantity of energy required to create tonnes of antimatter is so tremendous that it could never be produced in Earth-based factories. The only energy supply able to match these needs is the Sun. Gigantic solar panels (each several hundred square kilometres in area) would have to be set up in space, probably somewhere near Mercury, where solar light is most abundant. Antiproton factories would have to be supplied for years to manufacture enough fuel for interstellar travel. In other words, before this wonderful energy can be released by antimatter, it must first be gathered together and expended. At best, antimatter represents an efficient but expensive means for storing energy, rather than a natural source.

Rocketless rockets

A conventional rocket, considered as one which transports its own fuel and energy supply (on top of its payload), would appear to be incapable of opening the way to interstellar travel. Known energy sources, such as nuclear fission and fusion, are just not efficient enough, so that huge amounts of fuel are required. The most efficient source is antimatter, where relatively little fuel is needed. But it raises a different problem: the amount of antimatter needed, even for modest missions, far exceeds our current production capacity.

The situation seems hopeless, unless we take human ingenuity into account. For human willpower rarely admits defeat in the face of nature. A great many different ways have been explored by those who dream of deep space travel, with only modest success as yet. The common feature of all these projects is that they turn away from conventional rocket design. Interstellar spaceships are designed with no propellant and no fuel on board, and even without any motor! In other words, they are rocketless rockets.

The first to conceive of propulsion through space without rockets would seem to have been the Soviet astronautic pioneers Konstantin Tsiolkovsky and Fridrikh Tsander. In the 1920s, they suggested that the pressure of light from the Sun might exert sufficient push on a large sail to power travel within the Solar System. At the end of the 1950s, the American physicist Richard Garwin published the first technical article to discuss the idea of a solar sail ship. He drew attention to the fact that this method would cost almost nothing compared with other propulsion techniques under study at the time. Garwin illustrated its potential by giving a few concrete examples. The most impressive involved a return trip to Venus in less than one and a half months. A few years later, Arthur C. Clarke used this idea in his story *The Wind from the Sun*, describing a solar yacht race to the Moon.

The idea behind solar sailboats is fairly straightforward, appealing to a well known property of light. Photons are particles of light. When they strike a surface in their path, they exert a force on it. The resulting pressure is naturally very small, unless a very great number of photons is involved, for example by using a sufficiently extensive sail. In this case, the pressure can result in a significant force, quite capable of accelerating the sailboat to high speeds. We should emphasise the difference between the gentle breeze caused by solar photons, always blowing out radially with the same intensity at a given distance, and the blustery wind of high-energy particles (electrons, protons and atomic nuclei), whose intensity fluctuates considerably in time (notably, during solar flares, as we saw in Chapter 1). The pressure due to solar wind particles is significantly smaller than that caused by photons emitted from the Sun.

The surface of the sail must be highly reflective, for two reasons. Firstly, this reduces the risk of overheating by absorption of light energy, and secondly, it maximises the pressure obtained by allowing photons to bounce off. A silver sheet would be ideal, but somewhat costly! The most promising material seems to be aluminium. Its reflecting power can be made as high as 90%, so that nine out of ten photons will bounce off. In addition, the sail must obviously be made as thin as possible to keep its mass within reasonable limits. Indeed, the lower the mass per unit area, the higher will be the acceleration for given total area. Something rather like a spider's web comes to mind, with a very large area and extremely reduced thickness.

The idea of navigating the Solar System by sail power may seem rather paradoxical, since the Sun's photons always exert their pressure in the radial direction. It might be thought that we could only move away from the Sun using this technique. But this is not so. Just as we can navigate on Earth's surface with winds that always blow in the same direction, so we can move without restriction through the Solar System, thanks to the Sun's radial photon wind. In both cases, it is the sail orientation that makes it possible to change direction, and even to travel upwind, although the physical phenomenon at the root of this possibility is not the same. On the sea, it is the resistance of the keel to lateral motion that gives the desired effect. In space, the solar yacht navigates using a well-known property of the gravitational field. The speed of an object orbiting the Sun depends on its distance from the star. Changing this speed, we can modify the orbit. If we wish the object to move towards the Sun, we reduce its speed. By reorientating the sail, we can thus reduce its orbital speed and bring it nearer to the

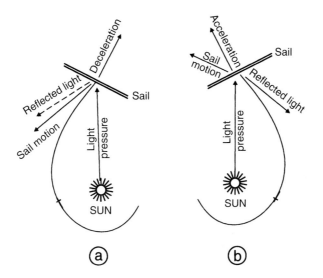

FIGURE 2.6   The basic principle of solar sailing within the Solar System. (a) The pressure due to solar light slows the sail on its trajectory. With less kinetic energy, it moves towards the Sun. (b) Solar light pressure accelerates the sail, which moves away from the Sun.

Sun; or we can accelerate it along the orbit and move it further away from the Sun (Fig. 2.6).

These ideas were investigated by NASA in the 1970s, in particular, at the Jet Propulsion Laboratory in California. Several technical problems were solved in a satisfactory way, at least on paper: choice of materials, how to deploy the sail in space, how to avoid damage by meteorite collisions, and so on. Possible missions included trips to Mars, Venus, a comet or an asteroid, expected to last several years. In each case, the sail had impressive dimensions, measuring several hundred metres across, but only a few micrometres thick. It weighed several tonnes, three or four times more than the payload it could transport.

As big as Red Square but lighter than an elephant, the solar sail would be a magnificent sight, billowing in a gentle breeze of solar photons as it swept through space to meet comet Halley. And what an elegant solution to the problem of interplanetary travel, avoiding all the waste of fuel and propellant inherent in rocket power. However, these projects never got the green light from NASA, even at the prototype stage. The true performance of solar sails is therefore unknown today, although they certainly look promising for travel within the Solar System.

Interstellar sails

At first sight, this method would not seem to apply to interstellar travel. The intensity of the Sun's light decreases as the square of the distance. As far out as Pluto, this intensity is already 1600 times less than it is at the distance of Earth's orbit. In interstellar space, it soon becomes negligible.

However, there are ways of overcoming this difficulty. Invention of the laser in 1960 made it possible to produce coherent light beams which could propagate over long distances without spread or significant loss of intensity. In 1962, Robert Forward, an engineer at the Hughes laboratory where the laser had been developed, suggested propelling an interstellar sail out to the nearest stars with the help of a superpowerful laser supplied by solar energy. Forward is probably the world's leading expert on the question of interstellar propulsion and has studied almost all the relevant problems. Some of his solutions are indeed original.

The advantages and disadvantages of the interstellar sail are well illustrated by reconsidering the Daedalus-type mission (unmanned flyby mission to a nearby star in less than 50 years, with a 450 tonne payload). The sail would be 30 kilometres across (almost as big as Greater London), but it would only weigh about thirty tonnes, being just 16 nanometres thick (16 billionths of a metre, equivalent to a few juxtaposed atoms). A power of a few terawatts would be needed, about as much as the present world energy consumption, to accelerate it to 60000 km/s ($0.2c$) after roughly thirty years. This power would be drawn from the Sun itself. The Sun's energy would supply a battery of laser emitters in orbit around Mercury. The advantage of this location is two-fold. Firstly, the Sun's light is more intense here, and secondly, Mercury's gravitational attraction would hold the emitters in place. Indeed, the emitted laser beams would be so powerful that there would be a considerable reaction force on the emitter, which must be balanced by the attraction of a massive body. Two hours later the laser beams would converge at a lens measuring 1000 kilometres across, positioned somewhere between Saturn and Uranus. The lens would be made up of concentric rings of ultrafine plastic film. Despite its spider's web appearance, it would weigh upwards of 500000 tonnes. A lens of this size could transmit a perfectly aligned laser beam, without the slightest divergence, for a distance of 40 light-years. Pushed for thirty years by this beam of blue–green light (optimal wavelengths for this kind of mission), the sail would reach 20% the speed of light. It could then reach the closest stellar systems within half a century.

Truly enormous constructions would be required in interplanetary space to set up this type of mission. However, these are merely an extrapolation on a very large scale of techniques that have already been mastered (at least, in theory!), and do not involve new technology. Just like the Orion and Daedalus projects, the interstellar sail ship shows that a mission to a nearby star, with a time scale of a few decades, is technically feasible.

The cost of such an enterprise should not be underestimated. It would certainly be of the same order as the production of antimatter. The current cost of energy produced by photovoltaic cells is around 2000 dollars per kilowatt. The terawatts needed for the mission described above would cost many hundreds of billions of dollars at today's prices. Such an ambitious aim could only be achieved in a few centuries from now.

It might appear that the method described here could not power a return journey to the stars. Unlike the solar sail, an interstellar sail cannot be slowed down by merely reorienting it, because the sail is not in orbit around the Sun (in the jargon of physics, its velocity has only a radial component and no tangential component). It was not until 1982, twenty years after proposing the idea of interstellar sail propulsion, that Forward realised how the very same laser beam could be used to decelerate the sail and bring it right back to its point of departure. His idea is extremely simple (at least on paper) and deserves mention. Indeed, he made use of it in his 1984 science fiction novel *The Flight of the Dragonfly*.

A manned mission could be sent 10 light-years from Earth, the distance to stars τ Ceti and ϵ Eridani, with the help of a multiple sail of total diameter 1000 kilometres (just a little bigger than the area of France). The system would include an inner sail 100 kilometres across

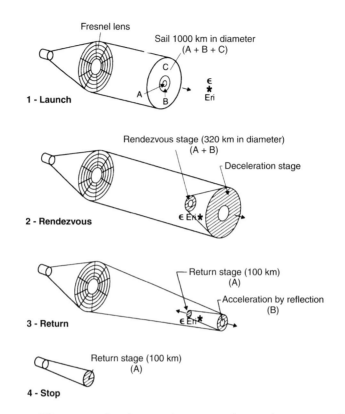

FIGURE 2.7   The outward and return journey to the nearby star ε Eridani (distance 10.7 light-years) using a sail pushed by a laser beam set up in the Solar System. (Adapted from Forward, R. (1988) *Future Magic*.)

(sail A, the only part involved in the return journey), surrounding the payload. It would also have two concentric ring-shaped sails measuring 320 and 1000 kilometres across (sails B and C, respectively), surrounding the first sail. Only one side of these sails would be made reflecting, by covering it with an aluminium film. The total sail mass (of the three sails) would be around 80000 tonnes, whilst the payload (spaceship, crew, provisions and exploration vehicles) would weigh 3000 tonnes.

Propelled along by a power of 43000 terawatts (just one ten billionth of the total power radiated by the Sun), the sail would be accelerated up to half the speed of light after eighteen months. The laser beam would then be turned off and the sail would continue at this cruise speed to reach ϵ Eridani after about twenty years. At 0.5 light-years from its destination, the sail would separate into two parts: the inner part, 320 kilometres across, comprising sails A and B and the payload; and the outer ring, sail C. The inner part of the sail would detach itself and turn on its own axis so as to direct its reflecting surface towards the outer sail. At the same time, the surface of ring C would be reconfigured in such a way that the inner part would now lie at its optical focal point.

The laser beam, reactivated ten years previously, would first reflect off the large outer ring, which would focus it on the inner part A + B. The latter would then begin to decelerate. One year later the inner part would come to a standstill in the region of ϵ Eridani, leaving the outer ring to speed off into space. Once the stellar system had been explored, the crew would begin to prepare for the return journey. The remaining sail would be further split into two parts. This time the reflecting surface of the new outer ring (sail B) would face the Sun. The innermost section (sail A) would be positioned at the focus of sail B. This part, measuring 100 kilometres across, would carry the spaceship and crew. Once again the laser beam from the Solar System would bounce off ring B and then off sail A, thereby accelerating the latter back towards the Sun. The return journey would last about the same time as the outward journey. A few months before arriving, the sail would swing round again so that the laser beam could reflect from its polished surface and slow it down. The whole mission could thus be accomplished in only forty-five years!

This may well look like a perfect solution: a manned return flight to a nearby star over a reasonable time scale, with neither fuel nor propellant to worry about! However, these figures should be put into perspective. The energy expenditure required for this type of mission exceeds Earth's present global production by a factor of several tens of thousands. Even assuming a steady increase by a few per cent per year, the levels needed here will not be reached for several centuries to come. And even then, it would probably be more profitable to use this colossal power to produce a few tonnes of antimatter, which could fuel an interstellar rocket with some degree of autonomy.

The main feature of sail propulsion, distinguishing it from other methods, is its dependence on the launch base. The mission could only succeed if the laser beam based in the Solar System were able to operate effectively over several decades, faithful to its commitment. If just one of its scheduled functions were to fail, the chances of a successful mission would fall to zero. If such a risky mission were to be manned, it could only interest astronauts with little desire ever to return to Earth!

Despite the difficulties, it seems that a miniature version of this propulsion scheme might well be realised in the not too distant future. The aim would be a reconnaissance mission to our closest stellar neighbour, α Centauri, consisting only of the twenty-year outward journey. The main feature of this project, devised by Robert Forward, is a drastic reduction in the energy expenditure and mass involved. It is heavily dependent on continued steady progress in miniaturising integrated circuits. Indeed, the mission's payload would not exceed 4 grams! Another original feature is that the sail would be propelled by a microwave beam. This is a form of electromagnetic energy which we are able to produce and transmit with extremely high efficiency. Compared with visible light, microwaves exhibit two major distinguishing features. Microwave beams begin to spread much earlier, implying that the sail would have to attain terminal speed rather close to the emitter. This in turn implies a much greater acceleration, whence the tiny mass of the payload. Furthermore, the sail can be perforated, allowing a significant reduction in its mass. This possibility, first put forward by Freeman Dyson in 1983, results from a well-known wave property of light. Photons cannot pass through a grid whose holes are smaller than their wavelength. It is for this reason that visible photons are reflected by a mirror, whilst gamma photons can pass through it. Microwaves have wavelengths of order one micron (one millionth of a metre). The sail could therefore be riddled with holes of this size, greatly reducing its mass.

On the basis of these ideas, Forward proposed the Starwisp project. The name suggests the weblike appearance of the sail: a spider's web one kilometre across and weighing only 16 grams! The 4 grams of intelligent microcircuits would be distributed along the ultrafine cables of the web, an exquisite challenge for the wizards of miniaturisation. This 20 gram microvehicle would be propelled by a 10 gigawatt microwave beam in orbit around Earth and supplied by solar energy. Under a steady acceleration of 115g (115 times the acceleration due to gravity at Earth's surface), the probe would reach a speed of 60 000 km/s in just a few days. It could thus arrive in the vicinity of α Centauri after around twenty years. During the few days of its encounter with our cosmic neighbour, it would then transmit observations back to Earth.

Ramjet, the ultimate starship

At the beginning of the 1960s, the concept of a 'miraculous' spaceship raised much enthusiasm amongst those who dreamed of interstellar travel. Imagine a starship that could accelerate steadily without any stock of fuel or propellant on board and which could even reach the most distant galaxies during the lifetime of its crew! This apparently crazy and very likely unfeasible idea went by the name of ramjet. At the time, it shook the world of astronautics.

In 1960, Robert Bussard, then working at the Los Alamos laboratory, published an article that revived dreams of rapid interstellar travel. It also renewed the enthusiasm of science fiction authors. Inspired by the idea of jet engines on aeroplanes, Bussard imagined a spaceship that would collect its fuel en route. In fact, interstellar space is not completely empty. It is filled with an extremely tenuous gas, mainly composed of hydrogen. The starship would collect and burn this hydrogen in its thermonuclear fusion motor, using the energy released to accelerate ejected combustion products. Without going into the details, Bussard assessed the potential of such a vehicle in an approximate way. Given the density of the interstellar medium, roughly one proton per cubic centimetre, and assuming that the reactor would burn hydrogen with 100% efficiency, he found that a 1000 tonne vehicle could accelerate at 1g for as long as it went on encountering fuel in its path. Starting out with a low speed of around 10 km/s (achieved by use of conventional rocket propulsion), it could reach 0.9c (270 000 km/s) after one year. It could then continue to accelerate, going past 0.99c, then 0.999c, and so on.

At such high speeds, the effects of special relativity would need to be taken into account. Ever since it was formulated by Einstein in 1905, this theory has been confirmed in the laboratory on many occasions. However, its consequences are an endless source of amazement, being so far-removed from our everyday experience. According to this theory, viewed from a given reference frame, the time in a moving system flows

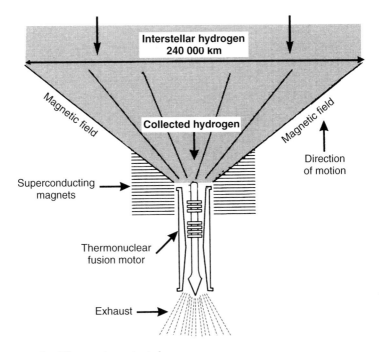

FIGURE 2.8   The ramjet principle.

more slowly than in one which is stationary with respect to that frame. This time dilation effect increases with the speed of the moving system, reaching quite impressive values when the speed approaches the speed of light. As an example, time elapses at only half the rate on board a rocket moving at 260000 km/s (0.86c), at only one tenth the rate when it moves at 0.995c, and at only one hundredth the rate when it has a speed of 0.99995c. This astonishing effect has given rise to many paradoxes. The most famous is undoubtedly the so-called twin paradox, illustrated in Fig. 2.9. No physical theory has caused greater disbelief and fascination amongst the general public in the twentieth century.

In an article published in 1963, American astronomer Carl Sagan set out the theoretical performance of a spaceship capable of accelerating continuously at 1g. After 3 years (time measured on board), the spaceship would reach α Centauri, 4.4 light-years from Earth. A year later it would be flying by ε Eridani, 11 light-years away. After 11 years the spaceship would enter the Pleiades cluster, 400 light-years away, and

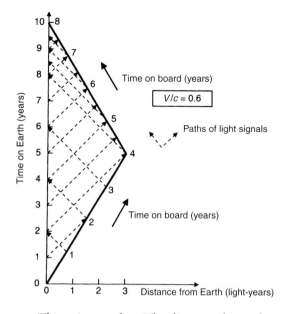

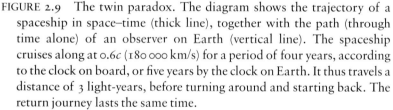

FIGURE 2.9   The twin paradox. The diagram shows the trajectory of a spaceship in space–time (thick line), together with the path (through time alone) of an observer on Earth (vertical line). The spaceship cruises along at 0.6c (180 000 km/s) for a period of four years, according to the clock on board, or five years by the clock on Earth. It thus travels a distance of 3 light-years, before turning around and starting back. The return journey lasts the same time.

   The Earth-based observer and the traveller exchange light signals each year (as indicated by their own clocks). Given the increasing distance between them, they receive the first signals 2 years after departure, and the second, 2 years after that (still with respect to their own clocks). The rate at which signals are sent is maintained throughout the return journey, but the change of direction leads to a shorter interval between the reception of two successive signals. Hence, during the outward journey, the spaceship receives only two signals, whereas it receives eight during the return journey (the last one at the instant of its arrival). Even though the two observers send signals at the same rate, the figure shows that the spaceship receives ten, whilst the observer on Earth receives only eight. The latter concludes that 8 years have gone by on board the spaceship, for 10 years elapsed on Earth. Time therefore flows 1.25 times faster for the Earth-based observer. For speeds close to c, these figures become all the more impressive. (Adapted from Mallove, E. and Matloff, E. (1989) *The Star Flight Handbook*.)

just 20 years after launch, it would be cruising through the centre of our Galaxy, 30 000 light-years away. Still accelerating, the spaceship would need only a further ten years to reach the Andromeda galaxy, 2 000 000 light-years from Earth. Finally, 15 years after that (and 45 after its departure), the spaceship would have reached the most distant galaxies, at the very limit of the observable Universe. Naturally, billions of years would have elapsed on Earth. Indeed, our planet would have long since become uninhabitable owing to the death of our Sun (see Chapter 3.). These amazing conclusions, based on relativistic physics, appeared in 1966 in the bestseller *Intelligent Life in the Universe*, cowritten by Carl Sagan and the Russian physicist Iossif Chklovski. The ramjet idea became synonymous with interstellar flight almost overnight and was henceforth widely exploited by science fiction writers. The best known story is probably *Tau Zero*, written by Poul Anderson in 1970. The ramjet *Leonora Christine* accidentally loses its decelerator during an encounter with an interstellar dust cloud. It is then condemned to accelerate endlessly across space whilst approaching ever closer to the speed of light. Thanks to relativistic time dilation, the ship's crew find themselves projected far into the future where they become witness to the dissolution and death of the Universe.

The basic idea behind the ramjet is simple. The faster the spaceship moves, the more hydrogen it can encounter per second, just as we are more quickly soaked when running through the rain than if we walk, intercepting more raindrops per unit time in the first case. Having ever more fuel available each second, energy production rises and the spaceship can steadily increase speed.

Detailed analysis of the ramjet soon showed that the idea was too good to be true. The fundamental problems raised are quite simply extraordinary. By comparison, the other projects for deep space travel considered so far do not look more difficult than a manned mission to Mars today.

In order to achieve a constant acceleration of 1g, the ramjet must collect hydrogen over a surrounding area of several tens of thousands of square kilometres. A suction device of this size would be unthinkable, so Bussard suggested setting up a gigantic magnetic field, as big as Earth! This field would channel hydrogen into the input nozzle, always assuming that the interstellar medium were ionised, since only electrically charged particles are affected by a magnetic field. However, the

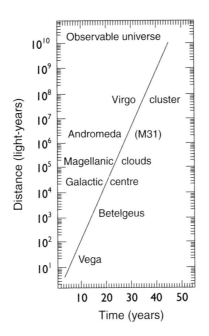

FIGURE 2.10 The theoretical performance of an interstellar spaceship able to accelerate steadily at 1g (acceleration due to gravity at Earth's surface). The time needed to reach the various objects shown is the time as measured on board the spaceship.

interstellar medium is mainly composed of electrically neutral atoms and molecules, which remain unaffected by electromagnetic fields. Only those regions in the neighbourhood of hot stars are actually occupied by ionised gases, because energetic stellar radiation can break atomic bonds. The ramjet engineers suggested a way of overcoming this first difficulty: a powerful laser beam could ionise the interstellar medium over tens of thousands of kilometres downstream of the spaceship. The second difficulty would be to attract particles from such an enormous region. Quite phenomenal magnetic fields would be required, so great that they would cause even the most powerful magnets to explode. Using the strongest superconducting magnets designed to date (which produce fields of around 1000 tesla), only one particle in every billion would actually be swept up.

In addition to this, hydrogen fusion is an extremely slow reaction. It may well allow the Sun to shine for billions of years, but there is no way that it could fuel a rocket engine. Deuterium would be a far more suitable fuel, with its much quicker reactions. It is for this very reason that it features in all present applications of thermonuclear fusion, whether they be explosive or controlled. But deuterium is a hundred thousand times less abundant than hydrogen in the interstellar medium, a fact which would do nothing to simplify the problem of gathering sufficient quantities to fuel the ramjet. A further possibility is to use isotopes of carbon and nitrogen as catalysts in the hydrogen fusion reaction. Indeed, this process occurs in massive stars and is extremely rapid (although still slower than deuterium fusion). A reactor capable of accelerating a 1000 tonne vehicle at 1g via this process would have to supply a power of 10000 terawatts, a thousand times more than the entire power generated by our civilisation today.

Having gathered the fuel, it must then be made to burn. But how could it be burnt whilst entering the motor at speeds approaching the speed of light? It would first have to be decelerated. However, this leads to an even greater problem. You cannot decelerate something without somehow absorbing the impact of the collision (even if it is inflicted 'gently' through the effects of a magnetic field). The ramjet would be considerably slowed down by having first to decelerate its own fuel. Some have even suggested that the ramjet would be better used as a brake for interstellar rockets!

The enormous technical obstacles involved in the ramjet have greatly damped initial enthusiasm towards the Bussard project. Several modifications have been put forward to render the idea more realistic, but they have all involved a drastic reduction in its miraculous capabilities. It would be pointless to list them all here, since none of them would appear to be more interesting than the other ideas presented previously. However, the notion of a rocketless rocket that could collect its own fuel along the way is interesting enough not to be neglected (*se non è vero, è ben trovato!*). After all, it is the only project existing today which offers the possibility of interstellar travel at relativistic speeds (at least, in theory). Tomorrow's engineers may succeed in overcoming the formidable difficulties posed by the ramjet. Intrepid scientists could then use it, not only to reach distant galaxies, but also to travel to the limits of the future, rather like the crew of *Leonora Christine*.

Relativistic visions

No human being, nor any measurement instrument, no matter how small, has ever travelled at relativistic speeds, that is, at speeds approaching the speed of light. Only elementary particles have been accelerated to such high speeds in our Earth-based laboratories. It should therefore come as no surprise that certain aspects of relativistic flight are so unfamiliar to the general public.

How would the sky look to the ramjet crew of the future, assuming that it could be turned into reality, accelerating steadily across our Galaxy until it reached ultrarelativistic speeds? We might naively imagine the stars streaming away behind us at great speed. But this is not at all the spectacle lying in store for future relativistic travellers.

Nothing extraordinary happens until we reach about $0.5c$ (half the speed of light). The stars ahead of us just appear to be a bit brighter than usual. Then, as we approach $0.9c$, the general appearance of the sky begins to change. Stars in front of the spaceship maintain their place in the sky whilst those located laterally (to 'left' and 'right', 'above' and 'below', although these notions have little meaning in space) seem to slide past in the forward direction rather than towards the back! The approaching field of view appears to fill up with more and more stars, whilst the lateral fields gradually empty out. At the same time, the stars are becoming brighter and brighter. At $0.9c$, the most brilliant amongst them are rivalling with planet Venus, the brightest object in the night sky after the Moon.

This brightening effect does not continue indefinitely. In fact it is accompanied by a gradual change in the colour of the stars. Blue stars are first to reach maximum brightness and then disappear from the cosmic scene, whilst the yellow, and then the red stars slowly turn blue, gradually brightening, before eventually disappearing in their turn. All the while, more and more stars are coming into view, which have remained invisible up until now. Hundreds of thousands of stars can now be seen, compared with a few thousand visible to the naked eye from Earth. An ever smaller sky is carpeted to saturation with bright points of light. The field of view shrinks to a circle just a few degrees across, straight ahead of the spaceship. Elsewhere the sky is plunged into total darkness. Finally, at the very highest speeds, touching on the speed of light itself, individual stars vanish completely. The whole

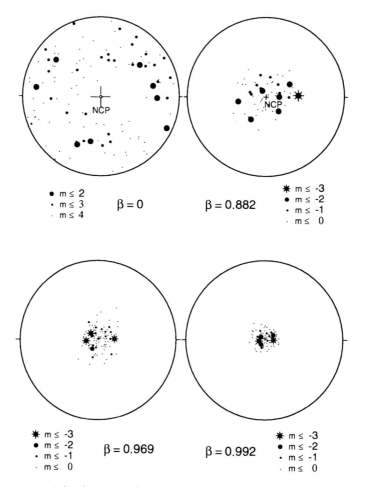

FIGURE 2.11 The sky as seen by the crew of an interstellar spaceship travelling at different speeds (expressed as a fraction of the speed of light $\beta = v/c$). The craft is moving towards the north pole of the celestial sphere (NCP). Symbols refer to the apparent magnitude (brightness) of the stars. The brightest are those with the most negative values. (Adapted from Sheldon, E. and Giles, R. (1983) *Journal of the British Interplanetary Society*.)

Universe is now concentrated into a minuscule region directly ahead, almost as bright as the surface of the Sun.

These images of the sky as perceived from a relativistically moving spaceship were produced by computer simulations and are due to optical effects, some of which are well known. For example, the changing colours of the stars is an instance of the Doppler–Fizeau effect: the frequency of the light we receive from a source approaching us is higher than the frequency of the same source at rest, and all the more so as the relative speed is great. (It makes no difference who is moving: the source, the observer or both.) The same effect explains why an ambulance siren seems to produce a higher note when approaching, and a lower one as it recedes.

Light emission from yellow stars, and thereafter red stars, appears to be shifted towards the blue. This is because blue corresponds to the highest frequencies of the electromagnetic spectrum that are visible to our eyes. At the same time, the blue stars eventually disappear, because their light emission is shifted into the ultraviolet, invisible to our eyes. Small stars, which are by far the most numerous, tend to emit the main part of their energy in the infrared, which is also invisible to our eyes. However, when they move towards us at sufficiently great speeds, their emission is shifted into the visible part of the spectrum.

A direct consequence of this effect is that the source appears brighter than it really is. Photons corresponding to higher frequencies carry more energy than those at lower frequencies. This explains the increased brightness of all the stars until they vanish into the ultraviolet.

At the very highest speeds, the brightness of all the stars merges with the light emission of the sky itself. In fact, the Universe is filled with 'cold' radiation, a relic of the hot primordial phase known as the Big Bang (Chapter 4). The frequencies of this radiation are a thousand times smaller than those of visible light. The whole Universe emits radio waves, invisible to our eyes, but detected by radiotelescopes since 1965. Photons making up this radiation are thousands of times more common than those emitted by all the stars put together. When the Doppler–Fizeau effect makes them visible, they effectively mask all other emission from the sky.

However, the most spectacular effect is the shrinking of the visible sky, due to light aberration. This effect is well known to astronomers. It is analogous to the way vertically falling rain traces out slanting paths on the side windows of a car when it is moving. The faster the car is going, the more the paths deviate from the vertical, giving the impression that their source is somewhere in front of the vehicle. Hence, the greater the speed of the relativistic spaceship, the more the stars (including those located behind the vehicle) appear to gather on its forward path, emptying out an ever wider region of the sky on either side. The astronauts might well feel that they are gradually slipping away from their familiar universe, and moving into an empty one, connected only by this shrinking umbilical cord of light.

The computer on the spaceship can certainly restore the Universe to its original proportions, as it would appear to a stationary observer at the instantaneous position of the vehicle, providing that it has previously stored the coordinates of most of the stars in its memory. In any case, these coordinates would have to be compared with the apparent coordinates of the stars, measured by instruments on board, in order to establish the speed of the spaceship, taking into account the effects of light aberration. However, at ultrarelativistic speeds, no star position can be measured. Each object loses its identity as it merges into the uniform glare of the cosmological background. Since navigation is no longer possible in these conditions, the spaceship can only follow some pre-established trajectory. Any change of path requires a deceleration to lower speeds, at which point the Universe will appear to open out like a rose in summer, reassuming little by little its former glory.

The hazards of relativistic travel

Relativistic travel is not without its risks, going far beyond the mere psychological consequences of visual effects on its crew, and related problems of navigation. In a previous section, we saw that the ramjet engine would be fuelled with hydrogen and deuterium gathered from the interstellar medium. This medium is also home to other chemical species, heavier than hydrogen. Some of these, like carbon, oxygen and silicon, bind together to form dust particles measuring typically a few millionths of a metre across and with masses in the neighbourhood of 10^{-16} grams. This dust is exceedingly tenuous. The average distance between two grains in a typical region of the interstellar medium is a few hundred metres. However, a spacecraft of diameter ten metres, moving at just $0.1c$ (30 000 km/s) could expect to intercept several thousand such grains every second.

Needless to say, dust grains constitute a major hazard to interstellar travel. Even though they are practically motionless relative to the interstellar medium, they certainly move relative to the spaceship, with a velocity equal (and opposite!) to the velocity of the vehicle. Travelling at $0.1c$, even such a tiny object carries a considerable amount of kinetic energy. Over the long years of an interstellar flight, continual bombardment by these projectiles might seriously damage the body of the spacecraft.

Paradoxically, damage would be considerably greater at modest speeds of order $0.1c$. In this case, material around the point of impact would vaporise, and outgassing would lead to gradual erosion of the walls of the spaceship. A graphite or beryllium shield several centimetres thick would be required to protect a manned spaceship from erosion during its several decades of interstellar flight at a cruise speed of $0.1c$. However, at ultrarelativistic speeds (very close to c), dust grains penetrate the outer walls of the spacecraft, but vaporised material immediately solidifies along their paths, cooled by contact with adjacent layers. Erosion would be negligible in this case.

There is no doubt that interstellar dust grains pose the greatest threat to the body of the starship. But cosmic rays represent an even more formidable challenge to the crew and electronic equipment on board. These particles, a mixture of protons, electrons and atomic nuclei, stream through space at relativistic speeds, accelerated by gigantic stellar explosions. In Chapter 1 we saw how dangerous these particles could be for interplanetary travellers. However, only the fastest amongst them (those moving above $0.5c$) manage to penetrate the inner Solar System. The rest are brought to a halt on the bounds of the heliosphere, about ten billion kilometres from the Sun, pushed back by the solar wind which emanates constantly from the surface of our star.

Once beyond the shield created by the heliosphere, the interstellar craft will have to face the full assortment of cosmic rays, and not only the most rapid amongst them. But even this is nothing compared with the threat posed by atomic nuclei in the interstellar medium. In relation to a relativistically moving spaceship, these normally inoffensive nuclei become high-energy cosmic rays, with considerable penetrating power. Electromagnetic fields must be used to repel these charged particles. In fact, the ramjet's magnetic net would channel them into the combustion chamber, as we have seen in an earlier discussion. But if this active shield were ever to weaken due to some system failure, the crew would be instantaneously roasted. Some more conventional passive protection must also be built in. According to estimates, a plate of lead about a metre thick would be needed to absorb the main part of the high-energy particles and thereby reduce the dose received by the crew of a relativistic spacecraft to acceptable levels. This would mean installing shields weighing several thousand tonnes, and hence significantly increasing the overall mass of the spaceship. Needless to say, fuel needs would rise in consequence.

Oddly enough, another aspect of relativistic space flights has escaped the notice of science fiction writers, with one exception. A simple calculation shows that a macroscopic object moving at relativistic speeds constitutes a bomb of quite extraordinary power, simply as a consequence of its great speed (see Fig. 2.12). The kinetic energy of a 1 kilogram mass moving at $0.3c$ is equivalent to the energy released by a 1 megatonne thermonuclear bomb when it explodes. (This is a typical element in today's panoply of nuclear weapons, fifty times more powerful than the bomb which destroyed Hiroshima.) At $0.99c$, the same mass is equivalent in kinetic energy to a 100 megatonne bomb, the most powerful weapon yet tested. The kinetic energy of a Daedalus-type starship (of mass 450 tonnes, moving at $0.15c$) is on a par with 120000 megatonnes of TNT, about twelve times greater than all the world's nuclear arsenals put together. It is comparable with the energy released by the impact of an asteroid measuring 1 kilometre across, which would cause a global disaster on Earth (see Chapter 3). The kinetic energy of a 10000 tonne ramjet moving at $0.99c$ is ten thousand times greater again, exceeding even the energy of the collision which ended the reign of the dinosaurs on Earth, 65 million years ago.

The figures cited here show to what extent a relativistic spaceship can be considered as a genuine missile. Any collision with a celestial body, be it accidental or deliberate, would lead to a quite unimaginable catastrophe, which could transform the whole surface of a planet into an inferno. Needless to say, the amount of energy required to accelerate the craft to such high speeds is truly colossal and far exceeds the

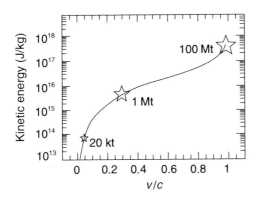

FIGURE 2.12   The kinetic energy (in joules) of a 1 kilogram mass for different values of its speed (expressed as a fraction of the speed of light). At a speed of 0.1*c*, its kinetic energy is equivalent to the energy released by the bomb which destroyed Hiroshima (20 kilotonnes of TNT). At 0.99*c*, it equals the energy of a 100 megatonne bomb, the most powerful weapon ever tested.

present capacity of our civilisation. For example, we could not accelerate Daedalus up to 0.15*c* even if we were to explode all the nuclear weapons in the world. However, if mankind does succeed one day in building these spaceships, they will have to be handled with the greatest caution. Indeed, an ultrarelativistic spaceship is more dangerous than a comet or asteroid of the same energy, simply because of its phenomenal speed. This point has been brought out in J. Pellegrino and G. Zembrowski's science fiction novel *The Killing Star*. When a relativistic object is detected, it is already too late, even to pray. An instant later, the object has already reached its target. As an example, the time elapsed between detection of a missile moving past Jupiter at 0.99*c* and its arrival on Earth is just a few minutes. In the case of a comet, we have several months' reprieve.

Sleeping astronaut... or immortal?

Reaching the stars, even the closest amongst them, would seem today to involve insurmountable difficulties. At present no starship project exists that is capable of transporting us to the stars in less than a human lifetime. But whilst science and technology work towards the goal of rapid and realistic transportation, we may wonder to what extent currently imposed limits on human lifespan are really a limiting factor for interstellar travel.

The three-cornered constraint: maximal speed less than the speed of light, enormous interstellar distances, and short human lifespan, has led science fiction writers to explore the idea of slow interstellar travel. This travel concerns either astronauts in a state of hibernation, or whole colonies living out several generations on board starship worlds. Such solutions very likely involve fewer technical difficulties with regard to means of propulsion. They nevertheless raise plenty of other questions, whether they be biological or physiological, in the case of hibernation, or psychological and sociological in the other case.

The idea of suspended animation to overcome the problem of lengthy interstellar journeys has some obvious strong points. Not only does it considerably reduce the amounts of air and provisions needed for the flight, but it also avoids potential psychological problems facing those destined to spend several decades in such a highly confined environment. Naturally, during this period the ship's computer would have to handle every aspect of the flight. This would include maintaining the crew's life support system and waking them upon arrival at their destination.

The idea has been marvellously illustrated in *2001: A Space Odyssey*, the cult film of Arthur C. Clarke and Stanley Kubrick. In order to cut down on provisions for a mission to Jupiter, three crew members are placed in hibernation. The other two command the spacecraft with the help of its resident computer, the famous HAL. The latter, suffering from an identity crisis, kills four members of the crew before being short-circuited by the fifth and only survivor. Clarke draws our attention to the fact that a blind faith in the machine involves certain risks. The key question is dramatically posed: can we justify trusting human life to machines during the long years of an interstellar journey?

It is not easy to answer this question today. Tremendous progress in the field of artificial intelligence and robotics makes it plausible that sooner or later the answer will be affirmative. After all, computers handle an increasing part of today's space missions. Since the complexity of these missions grows much faster than human ability to manage them, it is clear that the success of future interstellar projects rests primarily on the extent to which computers can be made to match the demand.

In contrast, we are a long way from achieving a state of hibernation in the human organism today, even if nature can supply us with several examples of this phenomenon. For certain wild animals are able to go into hibernation in order to survive the winter without food, lowering their body temperature and slowing down their metabolism. Unfortunately, humans do not have this ability, although some physiological effects do present a superficial similarity. Indeed, the medical journals contain several stories of people 'brought back to life' after their body temperature has fallen to very low levels. This sometimes happens to people in cold water after a shipwreck or trapped in the snow following an avalanche. When removed from their icy coffin several hours later, their heart has ceased to beat and their body temperature is down to 20 °C, well below the normal temperature of 37 °C. And yet they have been resuscitated, by warming up the vital organs with the injection of warm fluids. In a great many cases, such a return from the state of hypothermia to normality has been achieved without serious damage to the nervous system.

These cases would suggest that the brain can adapt to a restricted oxygen supply at sufficiently low temperatures, at least for a limited period of time. It might be thought that the human body could be put into a state of hibernation by lowering its temperature and reducing the metabolic rate and the rhythm of its other vital functions. However, research has shown that there are significant differences between hibernation and the survival of animals in a state of hypothermia. The most important is that animals can spontaneously emerge from the state of hibernation, whereas they cannot spontaneously recover from hypothermia. The ability to hibernate results from millions of years of natural evolution, whilst hypothermia is a brutally imposed external condition which organisms are not programmed to deal with. Various studies have shown that there are ways of assisting the organism confronted with hypothermia, such as high pressure oxygenation, glucose injections to avoid hypoglycaemia, control of blood electrolyte levels, and so on. Despite these possibilities, the present state of our knowledge would not justify any conclusion concerning artificial hibernation.

An extreme case of hibernation is known as suspended animation. In this case, bodily functions are not merely slowed down, but actually stopped altogether over a long period. The basis of this phenomenon is cooling to a temperature well below normal body temperature, in fact, right down below −130°C. At such low temperatures, molecular motions are so slow that any chemical reaction inside the cells is effectively suspended. Time appears to stop for the frozen cells.

Today a new branch of biology has come into existence: cryobiology. It has many applications, concerning both humans and animals. Frozen blood cells can be conserved for over a decade before being reused, whereas they would not survive more than a few weeks in conditions we normally think of as cold (i.e., around 0°C). Sperm banks are another example. It is commonplace today to conserve spermatozoids and ovules at −130°C for almost indefinite periods.

The main part of the cryobiologist's work consists in fighting the consequences of ice formation, which are anathema to any cell. As temperatures fall, water around the cells begins to turn into ice crystals. And water comprises on average 80% of the total volume of body tissue. Ice is lighter than liquid water and takes up a greater volume. This occurs to the detriment of any cells within it, which are first compressed and finally destroyed. In order to block this fatal process, the cryobiologist injects cryoprotective agents into the tissues before freezing them. The most effective seems to be glycerol, first identified as a cryoprotectant in 1948. At temperatures below −130°C, ice crystals are no longer able to form and cells remain intact. Experiments have shown that cooling must be carried out at some optimal rate for each different type of cell. For blood cells protected by glycerol, the rate is around a few hundred degrees centigrade per minute. Optimal temperature variation should also be applied when rewarming and resuscitating the cells.

Such methods have been used with some success on very small tissue samples (a few cubic centimetres at the most). Applying them to whole organs is quite another matter. In this case, there are large masses of tissues, composed of different types of cell. And as mentioned, each cell type requires its own optimal cooling rate. Even with the help of cryoprotectants, it is hard to envisage cooling all the parts of an organ, let alone a whole organism. Indeed, some cryobiologists claim that this will never be possible.

It is interesting to note that slow interstellar travel, in hibernation or suspended animation, raises a problem occurring in relativistic travel (described earlier). In both cases, the astronaut ages more slowly than those who remain on Earth. However, only relativistic astronauts can return to Earth during the lifetime of family and friends and suffer the psychological shock of seeing them in their old age. The sleeping astronauts know full well that on return from their slow trip (if indeed there is to be a return), no familiar face will be there to greet them.

Even more speculative than suspended animation is the idea of significantly extending human life expectancy. The world longevity record is at present held by Jeanne Calment who died in France in June 1997, at the age of 122. Certain animals, like turtles, can live up to two centuries, but none can reach the grand old age of Methuselah, who lived 969 years according to the Bible. Now for interstellar travel to become commonplace, even with slow means of transport, we would have to live at least a thousand years. A return trip to nearby stars at a few hundredths the speed of light would last several centuries, or several tenths the lifespan of a future Methuselah. This can be compared with expeditions undertaken by great explorers like Cook or Magellan, which lasted several years, and also represented a significant fraction of the crew's life spent on the high seas.

Such a spectacular extension of the human life expectancy is generally viewed with great scepticism by both scientists and general public alike. Of course, immortality is one of mankind's most ancient dreams, although recognised to be inaccessible. In every religion and mythology around the world, immortality is reserved for the gods. In fact, mortality is an essential feature of any living being. And yet, aging and death are merely biological processes. Even if they are poorly understood today, biologists of the future may succeed in deciphering and fighting against the genetic codes that control aging and death. If so, this will very likely involve a complete reorganisation of *Homo sapiens'* genetic heritage.

Clearly, if the elixir of immortal life (or even the elixir of longevity) were ever to be made available, the resulting demographic problems would create unprecedented social strife. Only a future civilisation with the entire resources of the Solar System at its disposal would be able to cope with such a situation, and then only for a few centuries. Even this civilisation must then look to the stars to guarantee its survival.

A hibernating or quasi-immortal crew sailing slowly across interstellar space might have to face another problem. As the centuries passed by, scientists back on Earth might discover some much quicker means of transport and prepare a mission to the same stellar system. On arrival, astronauts in the slow spaceship would find that the long years sacrificed for their trip had served no useful purpose. The objective would have been attained without them. Only the most desperate astronaut would accept this kind of trip.

It seems therefore that, quite apart from any progress in biology, such an approach to slow travel is unlikely ever to be undertaken. However, alternative methods for cruising slowly out to the stars have been put forward.

Voyaging worlds

It is difficult to imagine low-speed interstellar travel in a 'normal' spaceship, that is, in a small craft without comforts (except possibly for hibernating astronauts). A journey through space lasting several centuries would probably be a dull affair. Travellers, immortal or otherwise, would need as natural an environment as possible, in which they could live almost as they would on Earth. Such a mission would not be purely exploratory, unlike more rapidly propelled excursions. A permanent installation would probably be set up on arrival, implying a large crew. This type of trip also avoids the risk of frustration, in the event of an encounter with more rapidly moving explorers who have set off later but arrived first.

It is clear that very large space vehicles would be needed, able to carry a complete ecosystem and maintain it through almost perfect recycling capabilities. In the literature, this type of spaceship, a genuine scaled down version of Earth itself, is known as a 'world ship' or 'space ark'. Several generations would follow one another on board the craft over the centuries and millennia of the journey, hence explaining its other science fiction appellation, the 'generation starship'.

The first to formulate this idea explicitly would appear once again to be the father of astronautics, Konstantin Tsiolkovsky, in an article published in 1926. The English physicist John Desmond Bernal also considered it in his book entitled *The World, the Flesh and the Devil*, in 1927. Bernal was attempting to explore the long-term future of the human species. In his view, this future was characterised by the struggle between three traditional enemies of progress: the world, symbolising natural disaster; the flesh, symbolising enemies of the human body (disease, senility and death); and the devil, symbolising dark and irrational forces in the human soul (madness, jealousy, greed, and so on). Bernal pushed his vision of the future a long way, even anticipating some later discoveries in modern biology and physics. In particular, he envisaged human expansion across the Galaxy by means of gigantic constructions transporting thousands of passengers.

Bernal described a spherical world ship, 16 kilometres in diameter. (Since aerodynamics plays no role in empty space, the sphere provides the optimal volume per unit area of the outer surface. In other words, it yields the best inhabitable volume for a given mass of heavy materials used in constructing the hull.) The sphere is to be built using material gathered from asteroids, small natural satellites and other Solar System debris. The hull is made from an exceptionally strong material, able to absorb impacts but transparent so that electromagnetic radiation can pass through when the ship approaches some source of light. Radiated energy is absorbed by a fluid circulating within the hull. This fluid has similar properties to chlorophyll and can synthesise organic molecules from carbon dioxide gas. Raw materials such as carbon, oxygen and ice are stored in an inner layer about 400 metres thick. Finally, the inhabitable area inside covers about 2000 square kilometres, with a population of about 30000 passengers.

Bernal's description already contains all the main ingredients of the world ship, later to be developed by other authors. Most of these projects opt for cylindrical spaceships with masses varying between a few million and a few hundred billion tonnes. They are generally propelled by controlled thermonuclear fusion, with fuel extracted from the giant planets. Their occupants, numbering anywhere between a few hundred and a few hundred thousand, live in artificial gravity induced by rotation of the cylinder. Many of these ingredients (apart from the means of propulsion) were taken up by Gerard O'Neill and his students in their projects for space colonies around Earth (see Chapter 1).

The most detailed schemes were put forward by Alan Bond, leader of the Daedalus project, and his colleague Anthony Martin. They considered space arks big enough to reproduce a natural environment close to the one we enjoy here on Earth. One of their projects even contained an artificial lake. Their gigantic cylindrical spaceships would be 12 to 20 kilometres in diameter and as much as 200 kilometres long. The inhabitable area would be several thousand square kilometres (bigger than the Greater London area) and could provide shelter for three hundred thousand passengers. The hull of the vehicle would be made of steel plating several metres thick and would weigh several billion tonnes. This thickness would give the vehicle structural stability and would also protect passengers from cosmic rays, reducing exposure to an acceptable level.

Over half the length of the cylinder, the hull would be carpeted with a layer of regolith (the 'ground'), the remaining volume being occupied by an atmosphere. The other half would contain fuel in liquid form, ten times more massive than the body of the vehicle itself. Just as in O'Neill's space colonies, artificial gravity would be created in living areas by spinning the cylinder about its axis at a rate of once every five minutes, closely simulating gravitational effects on Earth's surface. For reasons of stability, the two halves of the cylinder, containing living environment and fuel respectively, would spin in opposite directions.

The power required to propel such space leviathans would be quite colossal. Engine power in the project devised by Bond and Martin can rise to several million terawatts, hundreds of thousands of times greater than the total power production of our civilisation today. This power is produced by deuterium fusion, as proposed for the Daedalus project. A hundred tonnes of deuterium would be detonated every second, releasing an energy equivalent to 2000 megatonnes of TNT (about 20% of today's world nuclear arsenal). Despite this huge power output, the world ship's acceleration would be almost imperceptible, just one thousandth of the acceleration due to gravity at the surface of Earth, because of its phenomenal mass. After running for fifty years, the engines would have pushed the space ark to its cruise speed of 1500 km/s or $0.005c$, so that it could reach the α Centauri system in about eight centuries.

Bond and Martin's gigantic world ship, christened Mark-2 by its inventors, bears a strong resemblance to one of science fiction's best known creations. This is Rama, the extraterrestrial craft in Arthur C. Clarke's famous tetralogy. In *Rendezvous with Rama*, first of the four volumes, the vehicle enters the Solar System in the year 2130 AD. It measures 20 kilometres in diameter and 50 kilometres in length. The

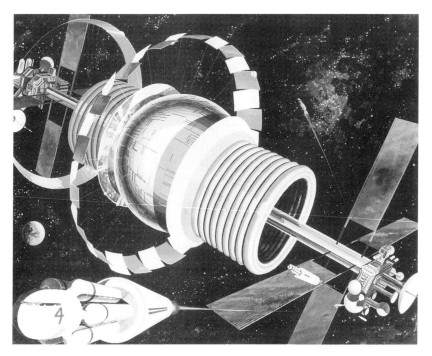

FIGURE 2.13   The Bernal sphere. The design was made in the framework
of the 1976 NASA Study on Space Manufacturing, concerning Space
Colonies in near-Earth space. The inhabitants would live in the inner
surface of a large sphere, nearly a mile in circumference, rotating to
provide them with gravity comparable to that on Earth. Their habitat
would be fully shielded against cosmic rays and solar flares by a non-
rotating spherical shell. The waste heat of the habitat would be radiated
away into the cold of outer space by the flat surfaces shown at each end
of the Bernal sphere. This type of design, augmented by a propulsion
scheme, could be adopted for Interstellar Arks. (Courtesy of NASA
Ames Home Page.)

hull is 500 metres thick and it has a total mass of 10 000 billion tonnes.
No propulsion system is described, which may leave the curious reader
feeling somewhat dissatisfied. An exploratory mission organised by
Earthlings discovers all sorts of strange life forms and robots inside
Rama, but no sign of the intelligent beings which built it. Without
attempting any form of contact with the Earthlings, the craft leaves the

FIGURE 2.14   A spaceship arriving in a distant solar system.

Solar System as silently as it arrived. It is a manifestation of Clarke's genius that this first contact occurs without real exchange between humanity and the extraterrestrials. (Needless to say, contact is established in the further three volumes of the set.)

It is not obvious that artificial structures as enormous as Mark-2 and Rama will ever be built. Apart from any purely technical considerations, it does seem rather a waste to procure only a few thousand square kilometres of living space from a cylinder 200 kilometres long! It would appear more natural to construct a world ship by converting the inside of an asteroid in the way described in Chapter 1.

Space ark sociology

The most interesting aspect of the space ark is perhaps not the technical challenge involved in designing an artificial cylinder, excavating an asteroid, devising a thrust system, and so on. Both science fiction writers and the general public are particularly fascinated by the sociological dimensions of such a venture. Could it not be the long-sought utopian society, this small and self-sufficient community of several thousand individuals lodged among the stars. The space ark is basically just a modern remake of the Fourierist phalanstery, revised and corrected by twentieth century science and twenty-second century technology. But the problems inherent in this kind of utopian society are well known: lack of contact with the outside world and lack of stimulus from new challenges, leading to withdrawal and stagnation, not to mention the rigid social organisation needed to preserve the community's fragile existence during the endless journey across space.

American writer Robert Heinlein was the first of the science fiction genre to explore these ideas in his novel *Universe*, in 1941. Harry Harrison's *Captive Universe* provides us with a marvellous illustration of sociological questions raised by world ships. At some undetermined point in the future, the asteroid Eros is transformed into a world ship to transport an Earthling colony to our closest stellar neighbour Proxima Centauri, where planets have been found. In order to maintain the vehicle in working order during the long centuries of the journey, a task requiring only a certain minimum intelligence and unlimited obedience, the crew are genetically manipulated. In addition, they are divided into two groups, constrained by strict laws to live separately throughout the trip. The two populations are only allowed to mix when they arrive at Proxima Centauri. Combination of the two groups is programmed to reveal the effects of a recessive intelligence-carrying gene. In this way the new 'smart' generation will be able to face the challenge of colonising the planetary system.

Despite all the precautions taken by the mission's designers, successive generations gradually become obsessed by the problem of running their craft, to such an extent that they eventually forget the objective of their mission. Thus, when they approach Proxima Centauri, they decide not to stop the spaceship, but carry on into interstellar space. Decades later, a couple finally manages to break the taboo separating the two populations and give birth to the first gifted child. The latter discovers the age-old and forgotten secrets of his world ship and persuades its other inhabitants that their objective actually lies behind them. Eros turns around to accomplish the mission it set out for so many centuries earlier.

A great many other writers have explored the theme of the generation starship. Swedish poet and Nobel prize winner Harry Martinson was even inspired to compose an epic poem, entitled *Aniara*. Generally speaking, science fiction authors tend to adopt a pessimistic stance. World ship missions only rarely fulfill the objectives of their designers, and more often than not, end in disaster. Such a pessimistic vision is fertile ground for dramatic intrigue, but is not necessarily very realistic. We can easily imagine these societies evolving in a perfectly stable manner, whilst manifesting all the vitality and creativity required of them.

One solution has been put forward regarding the problem of fossilisation, that threatens any closed society left to its own ends for many centuries. The idea is to launch a whole fleet of low-speed world ships in communication with one another. This suggestion first occurred in *The World, the Flesh and the Devil*, although for a slightly different reason. Bernal was concerned by the possibility of genetic degeneracy associated with reproduction within a limited population. He believed that interbreeding between the populations of the various world craft would deal with the problem. However, such a risk would appear to be negligible today. With sufficient understanding of our genetic code, this kind of threat could be effectively countered.

One further question, of an ethical nature this time, has been raised by Edward Regis Jr, professor of philosophy at Harvard University. Despite all the comforts which could be incorporated into them, these space arks will never constitute a perfect copy of the Earth environment. What right do astronauts have to condemn their children to this permanently enclosed life, where they will never witness the magic of a clear sky nor many other wonders of our blue planet? Regis concludes that this is a false problem, however. After all, Earth itself is far from being a paradise. Whole regions are affected by war, famine, epidemics or pollution. And yet children are born into this every day, eager for life. The generation starship would admittedly be better than a living hell, but could it ever be a paradise?

Nomads of interstellar space

So many difficulties are involved in building relativistic vehicles, apart from the problems raised for future astronauts by relativistic time dilation, that we may expect the Galaxy to be colonised rather by slow migration of world ships through interstellar space. Before undertaking such a voyage, human beings will certainly have colonised the rest of the Solar System, including the asteroid belt and the cometary Oort cloud described in Chapter 1. As time goes by, the inhabitants of these new worlds will feel less and less sentimental attachment to the mother planet. The Sun will be cold and faint in their dark sky and will be of little significance to the more distant of these colonies. Their energy needs will probably be covered by controlled thermonuclear fusion. Deuterium and helium-3 abound in the outer regions of the Solar System, as do all the essential raw materials such as water, carbon, nitrogen, oxygen and metals. These miniature worlds could one day become completely autonomous.

Gradual installation of new colonies in distant regions might eventually lead to shortages of raw materials and fuel. Under the pressure of overpopulation, some of the miniature worlds would probably choose to leave the Solar System and head for the stars. Their inhabitants would not be particularly prone to the psychological problems described in the last few sections, for they would be travelling through the same sky and in the same world that they had always lived in, rather than in some inhospitable vehicle. Their own artificial sun would provide them with far more light and heat than the Earthlings' Sun.

Under the icy light of distant stars, these worlds would slowly drift away from the Sun. It would take many decades to leave the cometary cloud surrounding our Solar System, which is tens of billions of kilometres from its guiding star. Then for long centuries, they would continue their journey across the immensity of interstellar space in the direction of a nearby star. They must first ensure a sufficient supply of raw materials, because they could not pin all their hopes on a chance encounter with some other body.

As they approached their chosen destination, the settlers' first concern would be to locate comets and asteroids, debris left over from the formation of the stellar system, in order to restock their supplies. Strangely enough, they would pay little attention to the planets that might be orbiting the new star. As we saw in Chapter 1, it is much easier in energy terms to exploit the resources of an asteroid than to overcome the gravitational well of a planet and transport materials from its surface into space. Only a planet offering similar climatic conditions to those prevailing on Earth would be of any interest to settlers, and such a planet remains an unlikely possibility. In the event of such a discovery, some would choose to colonise the planet's surface and begin a new planetary civilisation, just like the one made by their distant ancestors on Earth. But many other settlers would find the surface of the planet fragile and inhospitable, under the constant torment of bad weather and the threat of collisions with objects from space (asteroids and comets). They might well prefer to stay in their protective cocoon, orbiting the new star autonomously, in the way they have always been used to.

After a few more centuries, some settlers would decide to leave their stellar system and emigrate to another star several light-years away. The same scenario would repeat itself, maybe tens of thousands of times. Hopping from star to star, Earth's descendants would colonise nearby space, then more and more distant regions. Hundreds or thousands of centuries from now, an ever growing fleet of world ships would be streaming out across interstellar space.

Galactic civilisation

Three types of civilisation might emerge in this distant future. The first, rather conventional, would elect to inhabit the surfaces of planets resembling Earth with their own natural or artificial atmosphere, obtained by terraforming if necessary (see Chapter 1). A second type of civilisation would set itself up among the countless asteroids and comets that orbit around stars, using energy released by the associated star and raw materials so abundant in their vicinity. The third type of civilisation would prefer to venture forth across the vast expanse of interstellar space aboard well-stocked world ships. This mode of life would attract the most independent souls, opposed to control by some central power.

This vision of the distant future, already sketched out in the works of Tsiolkovsky and Bernal, may appear somewhat naive. But it remains as plausible as any other prospect we may imagine for the future of our species. In contrast, there is another vision which is dear to many science fiction writers and which seems wholly unrealistic. This concerns a central power imposing total control over regions as far apart as several hundred or even several thousand light-years. Such an interstellar empire would be impossible to hold together because it would take so long to communicate between its various components. Only superluminal travel could overcome this difficulty.

The best illustration of a galactic empire occurs in what is probably Isaac Asimov's most famous science fiction work, the series entitled *Foundation*. Somewhere around the year 10000 AD, a galactic empire of twenty-five million worlds inhabited by our descendants, is made possible by travel through hyperspace. The four volumes of Asimov's work describe the fall of the Galactic Empire (whose capital planet is Imperial Trantor), then the ensuing dark centuries of anarchy and decay. Two successive foundations, the last bastions of civilisation on outlying planets, attempt to bring the empire back to its former glory. Asimov was clearly inspired by the fall of the Roman empire, followed by the long centuries of the Dark Ages. His most interesting discovery is psychohistory, a hypothetical future science based on mathematics, history, psychology and sociology. Psychohistorical analysis leads to predictions concerning the behaviour of a large group of people, even a whole society, over thousands of years. Through this technique, scientists on Imperial Trantor actually predict the fall of their own empire. Since it is too late to sway the march of destiny, they set up a millenarian resurrection plan to be implemented by two foundations. Let us hope that psychohistory, or any other 'science' of this kind, will remain forever fiction. It could lead to disastrous effects, allowing some future Big Brother to control a society for many centuries.

A much more important question than the nature of the galactic empire is raised by the expansion of the human species into other stellar systems or interstellar space: how will humans evolve, living thousands of years in isolated environments so very different from one another? We cannot yet supply an answer to this question. The theory of evolution, originally devised by Charles Darwin and Alfred Wallace and then considerably developed by their successors, assumes that evolution essentially results from the interplay of two basic factors: random mutations of genetic material and environmental influences which favour certain species to the detriment of others. However, the theory will not necessarily apply to the future evolution of the human species, for the following reason. Unlike any other known species, *Homo sapiens* has developed a technological civilisation which effectively isolates it from environmental pressures. Rather than merely suffering the vicissitudes of the natural environment, people adapt it to meet their new needs. Our ultimate aim is to control our environment completely. If we do not succeed, then it will clearly be impossible to build world craft and consider disseminating our civilisation across the Galaxy. But controlling our environment will mean that, for the first time in the history of life, we are inhibiting the mechanism of natural selection. The human species may thus evolve without significant change, for as long as we are able to anticipate, even in the most hostile of environments.

These ideas are discussed by Freeman Dyson in his book *Infinite in All Directions*. This is how he views the future of the human species within the Galaxy:

> When life spreads out and diversifies in the universe, adapting itself to a spectrum of environments far wider than any one planet can encompass, the human species will one day find itself faced with the most momentous choice that we have had to make since the days when our ancestors came down from the trees in Africa and left their cousins the chimpanzees behind. We will have to choose, either to remain one species united by a common bodily shape as well as by a common history, or to let ouselves diversify as the other species of plants and animals will diversify. Shall we be forever one people, or shall we be a million intelligent species exploring diverse ways of living in a million different places across the galaxy? This is the great question which will soon be upon us. Fortunately, it is not the responsibility of this generation to answer it.

Colonisation of the Galaxy

Another important question is raised by the discussion in the last section, for reasons which will become clear in the rest of this chapter. If the human species ever masters the art of interstellar travel, either slow or rapid, how long will it take to expand across the whole Galaxy and settle in even the most distant regions?

It is clearly difficult to give a reasonable answer to this question. A lower limit is imposed by the size of the Galaxy, which measures almost a hundred thousand light-years in diameter (see Fig. 2.2). Even using relativistic vehicles, cruising at nearly the speed of light, hundreds of thousands of years would be needed just to cross the Milky Way. In slow-moving world ships, with speeds of a few thousandths the speed of light, the time required to cross the Galaxy becomes hundreds of times greater again, of the order of ten million years. The time required to colonise would clearly be greater still. But how much longer?

Some scientists have tried to apply quantitative methods to the problem of assessing how long it would take to colonise the whole Galaxy, inspired by demographic studies relating to populations of animals and people. The basic idea is simple. At any given place, the population of a species increases exponentially (with constant yearly growth rate), for as long as local environmental resources will allow it. When the limits of these resources are reached, the population must stabilise, either by reducing its growth rate to zero or by the departure of excess numbers towards new regions. Colonists settle nearby, wherever they find sufficient resources. Their own population also increases exponentially, with constant growth rate, until the system is once more saturated and the scenario repeats itself. A colonisation wave is thus created around the original location, and spreads outwards. Each new colony enters new and unoccupied territories. Propagation of such a colonisation wave front can be modelled mathematically.

Models of this kind have been applied with varying degrees of success to populations of animals and humans in different environments. Among the many cases of human propagation, the closest resemblance to a potential galactic colonisation is undoubtedly provided by the settling of islands in the Pacific. With the help of technological innovation, the adventurous peoples of the Pacific took several centuries to cross this immense ocean and discover virgin islands for settlement.

This great saga began in something like 3000 BC. Using simple boats, the coastal peoples of South-East Asia began to venture across the waters of the Philippine archipelago. After a few centuries, they settled on the southern coasts of New Guinea, and then on the many tiny islands in the Bismarck archipelago, slightly further east. Roughly fifteen centuries on, part of this population began the slow migration east, following a chain of more and more widely separated islands. Half a dozen generations later, this migration had brought them to the Fijian, Tongan and Samoan archipelagos, some 3000 kilometres further east. The explorers confronted the immensity of the ocean by inventing a new double-hulled boat, made by joining two canoes together. This was much more stable than their previous craft. It took several centuries to colonise these vast archipelagos, which constitute the home of the Polynesian culture today. Then in 100 or 200 BC, some of these Polynesians used an enlarged version of the double-hulled boat to make an even greater leap eastwards, reaching the Marquesas Islands and Tahiti, right in the heart of the Pacific. Finally, the last wave of colonisation took them to Hawaii in the north, New Zealand in the south and Easter Island in the East, four or five centuries later. In this way, about twenty centuries after leaving the Bismarck archipelago, almost all the Pacific islands had been colonised by the greatest navigators in the history of mankind.

The Polynesian adventure is certainly impressive, but we may well wonder whether there is really any similarity with colonisation of the Galaxy. Although distances between the Pacific islands are very great, no crossing would ever have required more than a few weeks, even on their rather primitive craft. Crossing an interstellar ocean on a time scale encompassing several generations is quite another story. Likewise, when it comes to galactic exploration, we may wonder whether population pressure will ever play such an important role as it did in colonisation of the Pacific. We could just as well imagine future societies capable of stabilising their populations before they begin to saturate their environment.

Nevertheless, the Pacific epic has inspired some scientists, like astronomers Carl Sagan and William Newman, and also Eric Jones, to apply the corresponding mathematical models to galactic colonisation. According to these models, the speed of the colonisation front depends on the mean distance D between two colonies, the time T_D needed to travel that distance, and the time T_C for the population of each new colony to reach saturation. Under these simple hypotheses, the propagation speed of the front is basically given by $V = D/(T_D + T_C)$. The longer the travel or saturation times, the slower the progress of the frontier. It is worth noting that the longer of the two times T_C and T_D tends to dominate the way the expansion proceeds and determine the speed of the colonisation front.

Interstellar distances are typically a few light-years. A slow-moving vehicle would take several millennia to cross them. If the time required to colonise a stellar system and prepare a new departure were less than one millennium, the colonisation front would move at a speed of about one light-year every thousand years. At this rate, it would take about a hundred million years for the wave of settlers to reach the most distant corners of the Galaxy. This may seem a long time, but it corresponds to only one hundredth the age of the Milky Way, estimated at more than ten billion years.

Von Neumann machines

The description of galactic colonisation presented in the last section assumes that successive generations of settlers will conserve the adventurous spirit of their ancestors over periods as long as ten million years. But it is easy to imagine a break occurring in the wave of colonisation after a certain time. The civilisation in some colonies may evolve in a very different direction which does not presuppose technological development and constant growth rate. It may, for example, evolve according to a spiritual dimension, with no inclination to expand across space. Other colonies may be completely destroyed, by use of weapons with a capacity for large-scale destruction, or by some unstoppable cosmic disaster, such as the explosion of a nearby star (as we shall see in the next chapter).

In 1960, radioastronomer Ronald Bracewell suggested that galactic colonisation would not be subject to the hazards described above if it could be carried out by robots, rather than human beings. Twenty years later, American mathematician Frank Tipler proposed using a special type of robot, able to do something which only living organisms can do today: self-replicate. These robots go by the name of von Neumann machines.

John von Neumann was one of the greatest mathematicians of this century. Of Hungarian origins, he emigrated to the USA in the 1930s and worked on the first computer and the first atomic bomb. In 1951, he devised an important mathematical model showing that it is possible to build a machine complex enough to produce an exact replica of itself. The self-replicating machine would consist of two main parts: the constructor and the constructing program. The constructor can transform basic materials and put them together according to instructions given by the program. If it is able to construct absolutely anything, it is given the name of universal constructor. The program contains all instructions needed to fabricate a perfect replica of the machine. To begin with, the constructor produces a copy of itself (a 'stupid' machine). Then, still following its instructions, it adds in an exact copy of the program. The result is identical to the first machine and can in its turn self-replicate (provided it finds the necessary materials in its vicinity).

The von Neumann machine is a purely theoretical concept today. We do not know how to make one and will probably not be in a position to do so for many centuries to come. One thing is clear, however. In complexity, it will far surpass any virus, and will be an entity lying somewhere between the living and the inert. Indeed, viruses use certain cells in the host organism to simplify their task. The von Neumann machine must manufacture all parts of its progeny from A to Z. According to some estimates its program will contain several tens of millions of instructions. Progress in the field of nanotechnology and molecular engineering indicate that it will weigh much less than original estimates of around a thousand tonnes.

Tipler believes that a sufficiently highly developed technological civilisation will be able to construct such self-replicating machines. It could then use them in a programme of long-term colonisation across the Galaxy. They would be much less demanding than human settlers. Indeed, machines could travel through interstellar space without needing breathable atmosphere or food. Once they arrive in the specified stellar system, they would use asteroidal material to manufacture several copies of themselves, together with the corresponding number of interstellar spaceships. From each stellar system, several machines would leave in the direction of slightly more distant destinations. The mother robots, once they had packed off their offspring, would begin the task of exploring their predestined stellar system, transmitting detailed data back to Earth.

In this way, the wave of self-replicating robots would gradually spread out across the whole Galaxy. Assuming that each robot sends out just two others in the direction of neighbouring stellar systems, one machine would be gravitating around each of the hundred billion stars in our Galaxy after just 36 generations. At this point, they must cease to multiply, otherwise all the heavy materials in the Galaxy would be absorbed into their construction within just a few more generations. The robots' program, something like their genetic code, must therefore include a specific instruction leading to the sterility of the thirty-seventh generation. This would have been introduced at the outset by their long-dead designers and transmitted from generation to generation until it came into effect.

The time required by these robots to colonise the Galaxy would depend critically on the speed of the interstellar spacecraft that transported them. The time used in building the next generation would be negligible in comparison. At a speed of $0.1c$, it would take as little as a few million years, whereas in low-speed vehicles, the time span would be more like a hundred million years. After this period, the descendants of those who designed the program on Earth would have complete and detailed knowledge of the whole of the Milky Way, without ever having set foot outside the Solar System. Even more extraordinary, the whole undertaking would not cost more than the price of building and launching the first machine!

The plurality of worlds debate

The scenarios discussed up to now would suggest that any civilisation which had mastered the problem of interstellar travel could colonise the whole Galaxy in a time span of order 100 million years at the very most. This is a small fraction of the age of the Galaxy.

It happens that our Sun arrived relatively late on the cosmic scene, only 4.5 billion years ago. Several generations of stars came before it during the 12-billion-year lifetime of our Galaxy. It may be that other forms of life were born around some of these stars, and some of these may have evolved into technological civilisations capable of undertaking interstellar travel. If this were the case, 'Where are they?' asked Italian physicist Enrico Fermi in 1950, summing up one of mankind's most ancient questions in a single, now famous phrase. Indeed, at least one of these advanced civilisations ought by now to have found its way here (because of the assumed rapidity of galactic colonisation), and left some trace of its presence. The absence of any trace of extraterrestrial civilisation in our Solar System does therefore throw doubt upon one or more points in the above arguments.

Among all the things that mankind has ever wondered about the Universe, perhaps the most fascinating question concerns the existence of an extraterrestrial life form or better still, an extraterrestrial civilisation. We are not in a position to answer the question today, although we can imagine the implications for the human species and its future in the Universe.

The debate about other worlds dates back at least twenty-five centuries to the thinkers of Ancient Greece. It is interesting to follow the history of this debate to see how arguments on either side have evolved with time, adapting themselves to the scientific developments of the day.

In ancient times, the word 'world' corresponded to the image that Aristotle and Ptolemy had made of the Universe: Earth at the centre, surrounded by the Moon, the Sun, the planets and distant stars. The plurality of worlds referred to the existence of several such universes, all completely independent and autonomous, with an inhabited Earth at the centre of each.

Most of the ancient thinkers (pre-Socratic Thales and Heraclitus, the Pythagoreans, Atomists Democritus and Leucippus, the Stoics Epicurus and Lucretius, and others too) believed in the plurality of worlds. Their reasoning was based partly on the fact that the Universe is vast (probably infinite) and partly on the principle of plenitude, which states that whatever is physically possible *must* exist somewhere. These arguments are best illustrated by a famous saying from Metrodorus, disciple of Epicurus: 'To consider the Earth as the only populated world in infinite space is as absurd as to assert that in an entire field sown with millet only one grain will grow.' Over the centuries, this argument has been taken up by those who believe in the existence of extraterrestrial life, without significant modification from its original form.

However, Plato and Aristotle, the two greatest thinkers of ancient times, were opposed to this idea. According to Aristotle's physics, there was only one Earth at the centre of a finite Universe. The space around it was divided into two parts: the sublunary, made from unstable combinations of four elements, earth, air, fire and water, imperfect and subject to change; and the astral, including the Sun, planets and distant stars, constituting a perfect, eternal and immutable world, made up of a single substance known as ether. Had there been any other earths beyond the sphere of stars, they would have mutually attracted one another and fallen towards the centre of the Universe. Furthermore, the empty space supposed to separate these worlds was not a notion occurring in Aristotle's physics.

The Aristotelian conception of the Universe was turned into a coherent system by the efforts of Alexandrian astronomer Claudius Ptolemy and dominated western thought for the next fifteen centuries, until the arrival of Copernicus. Thinkers in the early Middle Ages were also interested in the question of the plurality of worlds. They duly added theological arguments to support Aristotle's physical reasoning. Hence, in the fifth century AD, Saint Augustine rejected the idea of a plurality of worlds. In his view, the unique event represented by the Incarnation of Christ implied that there could be no other inhabited worlds. Needless to say, his arguments were strongly flavoured by Christianity's anthropocentric preconceptions, according to which the whole Universe had been created *for* mankind.

Whilst basically agreeing with Saint Augustine, Albertus Magnus could not help expressing certain doubts, still within the context of Christian theology. If God is omnipotent, why would he not have created other worlds? Saint Thomas Aquinas, a disciple of Albertus Magnus and founder of Scholastic philosophy, suggested an answer based on *reductio ad absurdum*. If God had made other worlds, he would either have made them all the same, or some of them different. (It would be hard to imagine another possibility, even for an omnipotent being.) The first case implies a particularly useless form of repetition, unbefitting of divine wisdom. The second implies that some of these worlds would be less perfect than others, contradicting the implicit perfection of the divine act. In this way, Saint Thomas concludes that there can only be one world: our own.

Such views seem rather unconvincing today, but they were sufficiently strongly held in the fifteenth century to have Giordano Bruno burnt at the stake. He had openly advocated the plurality of worlds, in an attempt to oppose the dogma which held that the Incarnation of Christ was unique. His heretical opinions were expressed when the Inquisition was already in full swing, and he is often considered as the first to be martyred in the name of science.

In the sixteenth century, Copernicus developed the heliocentric system and this, together with Galileo's observations, finally abolished the privileged status of planet Earth. Aristotle's arguments against the plurality of worlds fell to the same blow. At the end of the following century, however, Bernard Le Bovier de Fontenelle formulated another argument, much more significant than those of the scholastic thinkers. His book *Entretiens sur la Pluralité des mondes* (*Conversations on the*

Plurality of Worlds), which appeared in 1686, was a great success with the general public. It is often considered as the first popular science book, written in the form of dialogues between the author and a charming and ingenuous marchioness. The marchioness counters the author's assertion that 'intelligent beings exist in other worlds, such as the Moon' by the retort: 'If this were the case, the Moon's inhabitants would already have come to Earth.' Fontanelle can only argue that the time needed to master space travel is probably too long: 'If it is greater than 6000 years, we can understand why they have not yet been to visit us.' (In Fontenelle's day, the Universe was thought to be 6000 years old, after calculations by the Irish archbishop, Usher, based on biblical accounts.) In the remainder of this chapter we shall see how this argument, at the basis of Fermi's question, is reformulated in more modern terms.

In the eighteenth century, progress in astronomy brought the understanding that stars are just suns, rather like our own Sun, and this supported the view that an incalculable number of inhabited earths could exist in the Universe. The plurality of worlds was so well established in intellectual circles at the beginning of the nineteenth century that it was even used to dispute the dogma claiming uniqueness of the Incarnation of Christ, a complete reversal of the situation in Saint Augustine's day. In 1817, the Scottish theologian Thomas Chalmers wrote his *Discourses on the Christian Revelation viewed in connection with the Modern Astronomy*. His aim was to resolve the conflict. Without throwing doubt upon the existence of other worlds, he suggested that only the human species had committed original sin, requiring divine intervention for its salvation. The Incarnation of Christ had therefore occurred only once and the dogma was saved!

William Whewell, Master of Trinity College, Cambridge was the first to counter the plurality of worlds with a modern argument. In 1853, he drew attention to the fact that conditions on other planets in the Solar System are so different from those prevailing on Earth that no life form (at least, as we know it) could ever develop there. He also noted that, at the time, there was no proof of planets existing around other stars and that, during most of Earth's history, it had harboured no intelligent beings. In his view, the plurality of worlds was an argument developed, not on the basis of physical reasoning, but rather *against* all physical reason.

Whewell's opinions were not received favourably by his contemporaries. Faith in the plurality of worlds was further encouraged when, at the end of the nineteenth century, the American astronomer Percival Lowell announced the existence of artificially made 'canals' on the surface of Mars (see Chapter 1). In Lowell's view, these were a clear indication that Martians had undertaken a vast irrigation project on their planet. Although Lowell's observations were soon contradicted by his colleagues, they nevertheless had a strong influence on the general public. They thus inspired the founder of modern science fiction, Herbert G. Wells, to write his book *War of the Worlds*, probably the best known and most successful description of an extraterrestrial invasion of Earth.

At the beginning of the twentieth century, arguments inspired by the developing science of biology were added to the plurality of worlds debate. Alfred R. Wallace, cofounder of the theory of evolution, was the first to use this kind of argument *against* the idea of another intelligent life form in the Universe. In the 1905 edition of his book *Man's Place in Nature*, Wallace observed that mankind is the result of a sequence of unique and unpredictable events in the long evolutionary chain. The probability that this same sequence should occur elsewhere, even in an environment similar to the one provided by Earth, is practically negligible. The argument can also be applied to any intelligent life form.

Wallace's reasoning, adopted by many biologists, introduced a sense of history into the debate: a sequence of events may be of little importance when they are taken separately, but in combination, their effects are amplified over time to such a degree that the final result becomes completely unpredictable. It is interesting to note the similarity between this conception of history and modern chaos theory. According to this theory, which has mainly built up since the 1960s, the evolution of many physical systems is so sensitive to their initial conditions that it becomes impossible to predict their behaviour beyond a certain time horizon. In other words, virtually identical initial conditions can lead to completely different results.

The meaning of evolution deserves some reflection with regard to this point. In the traditional presentation of Darwinian evolution, emphasis is put on the progressive complexification of matter, as if there were some inevitable process here. The transition from bacteria to multicelled organisms, from fish to reptiles and from mammals to human beings is often considered as a deterministic, one-way movement. Along this path, natural selection rewards those organisms best able to adapt to their environment, by the survival of their lineage. However, as the American biologist Stephen J. Gould and others have pointed out, this view of evolution may be completely wrong. Natural selection is not the only factor determining the evolution of a species, and evolution does not always proceed in small steps. Major disasters have wiped out species which appeared otherwise to be earmarked for survival by natural selection. The best-known example is undoubtedly provided by the dinosaurs. After reigning on Earth for 130 million years (the same length as about 5 million human generations), the terrible lizards suddenly disappeared 65 million years ago. Their demise was probably due to a large asteroid that collided with Earth (see Chapter 3). Moreover, those creatures which survived the disasters were not always endowed with greater complexity than those that disappeared. Their comparative advantage was not always evident a priori. From this point of view, mammals owe their survival to simple good fortune and not to any particular superiority over the dinosaurs. Now at least four other major catastrophes have punctuated the 530 million years of multicellular life on our planet. We may therefore wonder whether the emergence of humans and intelligence occurring over the past few million years is not just the result of the purest good fortune.

These considerations are of the utmost importance for the question of other intelligent life forms in the Universe. Let us assume for example that planets exist in the Galaxy that are identical to our own, with conditions favouring prebiotic chemistry. Is it reasonable to think that a multicellular life form must eventually evolve? And what of more complex organisms, and intelligent beings? We are still a long way from answering these questions today. However, it would be quite astonishing if the emergence of life and its evolution towards more complex forms were as ineluctable a process as the evaporation of water at temperatures above 100 °C.

Where are they?

The question of the plurality of worlds has a long and fluctuating history, rich in new developments of a sometimes passionate nature.

Some arguments used in the past by supporters and opponents of the ETI hypothesis (ExtraTerrestrial Intelligence) are cause for amusement today. It is quite probable that some of our modern arguments will have the same effect on our descendants in a few decades or centuries to come.

Scientific study of ETI has a short history, going back only forty years. In an article published in 1959 in *Nature*, physicists Giuseppe Cocconi and Phil Morisson suggested that microwaves (high-frequency radio waves) are the best vector for interstellar communication. Not only do these waves penetrate the terrestrial atmosphere, but they can also pass through galactic dust and gas clouds. Visible photons, our traditional window upon the Universe, are absorbed by these clouds. Hence optical telescopes cannot see nearly as far as a radiotelescope across the disk of the Milky Way. In addition, radiotelescopes can survey the skies 24 hours a day, even in broad daylight or under thick cloud cover. As far as X-rays and γ rays, which are at the high-frequency end of the electromagnetic spectrum, are concerned, our atmosphere absorbs them and they cannot reach Earth's surface. (This is fortunate, since they are particularly harmful to living organisms.) Microwaves have another advantage: they carry little energy. This means that less energy would be expended in sending a message at these wavelengths. Cocconi and Morisson emphasised a third important point. Our Galaxy radiates little in the microwave region of the spectrum, compared with other radio frequencies. In other words, background noise would not interfere with communications.

These considerations opened up the modern era in the plurality of worlds debate, by initiating a scientific study of the problem, and the acronym ETI was born. The first to apply these ideas was Frank Drake, the young director of the Green Bank National Radioastronomy Observatory in the USA. He set up the first systematic search for extraterrestrial signals, called Ozma. The project was named after the queen of the imaginary country of Oz, a distant and inaccessible place, peopled with exotic creatures, in the story by Frank Baum. In 1960, the Green Bank radiotelescope spent two months looking for radio signals in the direction of two nearby stars, ϵ Eridani and τ Ceti, both about 12 light-years away. The negative results of this first experience did nothing to discourage ETI supporters. Dozens of other projects sprang into being, not only in the USA and the USSR, but also in

Canada, Australia, France and Holland. A few thousand hours spent listening to the heavens have revealed nothing to date. The searchers' initial optimism (reflected in the name of these projects: CETI, meaning Communication with ExtraTerrestrial Intelligence) gradually gave way to a more cautious attitude, at which point the project was renamed SETI, or Search for ExtraTerrestrial Intelligence.

Two results have come out of SETI up to the present time. One of them is probably durable and the other is probably temporary. Probes sent out to the distant confines of the Solar System have not indicated any life form in our immediate neighbourhood. Moreover, radiofrequency observation of the sky has not led to any detection of extraterrestrial signals. Given the size of the task, this result should come as no surprise. A much greater effort will be needed before any statistically significant conclusion can be drawn. However, even if we manage, over the next two centuries, to examine each of the hundred billion stars in our Galaxy across ten billion radio channels, what could we conclude from an absence of any artificial signal? We could only say that none of the supposed civilisations happened to be emitting in our direction. This hardly settles the question about whether ETI exist.

Apart from these search programmes, there is another observational fact whose importance is difficult to weigh up: the absence of any trace of extraterrestrial civilisation on our planet or elsewhere in the Solar System. This question, already raised by Fontenelle in his *Conversations on the Plurality of Worlds*, reappeared in its modern version around the middle of the twentieth century.

Towards the end of the 1940s, there came the first wave of reports concerning flying saucers and other UFOs (Unidentified Flying Objects), particularly in the USA. During a visit to the Los Alamos military laboratory in 1950, Italian physicist Enrico Fermi got into discussions on the subject with his colleagues. Among them was Edward Teller, future father of the American H-bomb. Everyone quickly agreed that the UFOs were unlikely to have extraterrestrial origins. The discussion moved on to the more general subject of extraterrestrial civilisations and interstellar travel. It was at this point that Fermi suddenly exclaimed: 'But where are they?' He went on to do a series of calculations to assess the probable number of civilisations in our Galaxy, concluding that they should already have visited us several times by now. In Fermi's view, the complete lack of evidence for such a visit did not necessarily imply that there were no extraterrestrials. It might just mean that insterstellar travel was impossible, or that technological civilisations were too short-lived, plagued by self-destruction upon discovering the secrets of the atom. (It should be remembered that the balance of terror between the USA and the USSR had just begun to hold sway at the time.)

This discussion between Fermi and Teller remained practically unknown for a long time. The question 'Where are they?', generally attributed to Fermi without further comment, first appeared in Sagan and Chklovksi's book *Intelligent Life in the Universe*, in 1966. In 1975, American astronomer Michael Hart independently rediscovered Fermi's arguments, without knowing about the discussion with Teller. His article came to the radical conclusion that the absence of extraterrestrials on Earth implies that we are the only technological civilisation in the Galaxy. He deduced that the continued search for radio signals would be a waste of time and money. Following this provocative article, Carl Sagan referred to this problem as the 'Fermi paradox'.

Hart's pessimistic conclusions began a period of passionate debate on the subject of ETI, notably in the USA. The controversy reached its climax in the early 1980s. In a series of articles, mathematician Frank Tipler observed that Fermi's paradox became even more paradoxical if it were assumed that one of the supposed civilisations were in a position to build self-replicating machines. As we saw earlier in this chapter, von Neumann machines could completely colonise the whole Galaxy in a relatively short time, quite independently of what was happening to the civilisation which had instigated their dissemination. The fact that there are no such robots in our Solar System is even more significant than the absence of any trace of extraterrestrial signals. In Tipler's view, it proves our technological superiority, if not our total solitude in the Galaxy.

Cosmic solitude?

Any paradox is based on at least one invalid assumption. The logical statement of Fermi's paradox is as follows:

A Our civilisation is not the only technological civilisation in the Galaxy.

B Our civilisation is in every way average, or typical. In particular, it is not the first to have appeared in the Galaxy, it is not the most technologically advanced, and it is not the only one seeking to explore the cosmos and communicate with other civilisations.

C Interstellar travel is not too difficult for civilisations slightly more advanced than our own. Some have mastered this kind of travel and undertaken a galactic colonisation programme, with or without self-replicating robots.

D Galactic colonisation is a relatively fast undertaking and could be achieved in less than a billion years. This represents only a small fraction of the age of the Milky Way.

If hypotheses A to D are valid, we can clearly deduce that 'They should be here'. The Fermi paradox applies. Supporters of ETI reject at least one of assumptions C and D, and some even go so far as to deny B, in order to save the key hypothesis A. In contrast, their opponents uphold the plausibility of C and D, whilst completely rejecting B. The most extreme even reject hypothesis A.

This is not the place to go into all the arguments for and against ETI made in the context of Fermi's paradox. Those arguments most often discussed do not refer to the physical aspect of the problem (the feasibility of interstellar travel and construction of self-replicating robots), but rather to its sociological features. Some argue that extraterrestrials would not even be interested in space travel, let alone expansion across the Galaxy. Their civilisation might have quickly turned to a spiritual way of life, occupying itself with contemplation and meditation; or it might have adopted the zero growth rate so dear to ecologists, making space colonisation unnecessary. Others, like Fermi himself, believe that a technological civilisation might be too short-lived, destroying itself before it could solve the problems of interstellar travel.

Such sociological arguments aim to reject hypotheses B and C. There is another class of sociological arguments, generally referred to as the cosmic zoo (or quarantine) hypothesis. According to this view, put forward by American astronomer John Ball in 1973, extraterrestrials have already visited our Solar System, either in the recent or distant past, but prefer to observe us from afar for various reasons. For example, they may consider us too primitive, they may not wish to interfere with our development, or again, they may fear our atomic weapons!

It should be noted that the first person to discuss these issues in the twentieth century seems to have been the Russian savant and father of astonautics Konstantin Tsiolkovsky. He put forward several possible explanations for the lack of evidence to support the claim that there is intelligent extraterrestrial life. For instance, he observed that Europeans discovered native Americans only many thousands of years after the beginning of their civilisation (Fontenelle's argument). But he preferred the so-called zoo hypothesis, which argues that 'we have been set aside as a reserve of intelligence in order to allow our species to evolve to perfection and thereby bring something unique to the cosmic community'.

All these sociological arguments share a common weak point. It is hard to believe that any one of them could apply to every single extra-terrestrial civilisation in the Galaxy. At least one hypothetical civilisa-tion ought to escape self-destruction, solve the problem of space travel and undertake a programme of galactic colonisation. The behaviour of animal species on Earth shows that they always go through an expansion phase, favoured by natural selection, just because this maxi-mises their chances of survival. What is more, at least one of these civ-ilisations ought to have overcome the taboo exhorting them to avoid all contact with our own civilisation. If none has come to this point, then assumption B is implicitly violated. For in this case, we would be the only ones seeking contact with other civilisations.

It is interesting to note that sociological arguments are generally invoked by those who advocate a search for radio signals. There is an obvious inconsistency in this position. Consider one of the first extra-terrestrial civilisations wishing to communicate with other forms of intelligence. It would be easy to show that, even in the most favourable case, the closest civilisation would be hundreds or thousands of light-years away. Consequently, no reply would come to their radio signals before several centuries had elapsed. In such conditions, it would seem more logical to invest in a space travel programme. If they explored neighbouring stellar systems using interstellar spacecraft, they would at least possess concrete information after a few centuries, even if there were no other civilisations in existence. A strategy based purely on radio emissions could go without results for thousands of years.

Explaining away Fermi's paradox by sociological arguments seems to me extremely questionable. It would be quite a different matter if there existed a sociological theory explaining why all civilisations must behave in this way. However, I doubt whether such a theory will ever be formulated. I also find it difficult to accept the physical argument invoked by Enrico Fermi in 1950 (during his discussion with Teller), and independently by British astrophysicist Fred Hoyle. In their view, interstellar travel is quite simply impossible. The speculations presented in this chapter are then merely a naive vision of reality, seriously underestimating the difficulties involved. Our species is therefore condemned to remain forever within the confines of the Solar System, until the death of our star. However, no physical law seems to forbid the accomplishment of such journeys. Difficulties are quantitative, rather than qualitative, and it seems unlikely that they will permanently block the way to deep space travel.

The most economical solution to Fermi's paradox consists in straightforwardly rejecting hypothesis A, as proposed by Hart and Tipler: our civilisation might just be the first technological civilisation to have appeared in the Galaxy. This solution is consistent with our present understanding of the theory of evolution, according to which the emergence of intelligence has been a rather improbable event. It is a significant fact that supporters of ETI are mainly astronomers, whilst biologists remain neutral, or openly hostile.

In Tipler's view, the main motivation for proponents of ETI is metaphysical: their hope is that extraterrestrial intervention will save us from ourselves. In his book *Broca's Brain*, Carl Sagan writes: 'It is possible that among the first contents of such a [radio] message may be detailed prescriptions for the avoidance of technological disaster, for a passage through adolescence to maturity.' In the introduction to his anthology *Interstellar Communication*, Canadian astrophysicist Alastair Cameron writes: 'Perhaps we shall also receive valuable lessons in the techniques of stable world government.' The leading light in the search for ETI radio signals, Frank Drake, expresses an almost religious desire in his eloquently titled article *On Hands and Knees in Search of Elysium*:

It is extremely likely that any civilization we detect would be more advanced than ours. Thus it would provide us with a glimpse of what our own future could be.... It is the immortals we will most likely discover.... An immortal civilization's best assurance of safety would be to make other societies immortal like themselves, rather than risk hazardous military adventures. Thus, we could expect them to spread actively the secrets of their immortality among the young, technically developing civilizations.

This optimism with regard to the potential benefits of an encounter with an extraterrestrial civilisation is not shared by all. Ever since H.G. Wells' *War of the Worlds*, the sombre image of a threat leading to slavery or the extermination of humanity has been far more widespread in science fiction literature. In his *Profiles of the Future*, Arthur C. Clarke was clearly influenced by that master of imaginative writing, Howard P. Lovecraft, when he wrote: 'The road to Lilliput is short, and it leads nowhere. But the road to Brobdingnag is another matter; we can see along it only a little way, as it winds outwards through the stars, and we cannot guess what strange travellers it carries. It may be well for our peace of mind if we never know.' Fortunately, in his other works, Clarke adopts a much less xenophobic position. At times, he even swings to the opposite extreme, thereby falling into line with Sagan and Drake.

Today the plurality of worlds is a more controversial issue than ever. Arguments on both sides ('It is unlikely that we are alone in the Universe' and 'Where are they?') are of a statistical kind. They are consequently of little import, for statistics cannot be based on the single case provided by life on Earth.

Detection of some inhabited planet, or better still, some extraterrestrial civilisation, would undoubtedly be one of the major landmarks in the history of mankind. On the other hand, non-detection of ETI signals, even after centuries of research, would never prove that there were no extraterrestrial civilisations. But it would be reason to prepare ourselves for a life of cosmic solitude.

STAR MAKERS

The only way of discovering the limits of the possible is to
venture a little way past them into the impossible.

Clarke's second law

Arthur C. Clarke, *Profiles of the Future.*

We will most probably be unable to reach even the nearest stars within
the next couple of centuries. And even if we do succeed, it is not certain
that we will find planets around them capable of supporting life.

This rather frustrating observation forces us to turn our gaze back
to the more familiar cosmic environment provided by the Solar System.
In the first chapter of this book, we saw that in the relatively near
future, that is, within the next few centuries at most, we could set up
bases on the Moon and Mars, and exploit asteroidal material in
Earth's suburbs. It is also conceivable that part of humanity will
choose to live in artificial space colonies, like the ones proposed by
Gerard O'Neill at the end of the 1960s, or 'natural' colonies inside hollowed out asteroids. And finally, certain planets, such as Mars, may be
terraformed to provide them with an inhabitable surface, although
such vast projects remain purely speculative at the present time.

All these futuristic visions assume that the human species will continue along the road to progress defined by increased energy consumption and material wealth (but not necessarily greater happiness!).
Adopting this hypothesis, we can try to imagine the very-long-term
future of mankind within the bounds of the Solar System. It is clear
that, sooner or later, we must come up against a limiting factor that is
rarely taken into account, namely the finiteness of resources contained
within any physical system.

We are already reasonably conscious that terrestrial resources are finite (although it turns out that the study carried out by the Club of Rome, and which sounded the alarm in 1972, was too pessimistic and even wrong about certain points). This has been so beautifully illustrated in the title of Albert Jacquard's book *Voici le temps du monde fini* (*Hail the Age of the Finite World*). For example, assuming steady growth of the world population, no matter how low the growth rate, the Solar System will be saturated within a short period relative to the present age of humanity. In the same way, it can be shown that with a 0.01% annual growth in world energy consumption, mankind would be using the equivalent of all the power radiated by the Sun within the relatively short period of three hundred millennia, a negligible lapse of time in astronomical terms (see Fig. 3.1).

This kind of extrapolation, so often used in the past, teaches us very little about the distant future. It shows how absurd it is to assume continued exponential growth, and highlights in a dramatic way the finiteness of any physical system. We may nevertheless pose the problem in different terms: how can a sufficiently advanced technological civilisation best exploit raw material and energy resources in its stellar system, in such a way as to maximise its population, its energy consumption and its own durability?

The Dyson sphere

In the 1960s, Freeman Dyson gave a tentative answer to this question. He was not really concerned about the long-term future of humanity, but rather by the possibility of detecting some technologically advanced extraterrestrial civilisation, even without its becoming aware that it was under observation.

In order to understand his argument, we must go back to the end of the 1950s. The idea of looking for extraterrestrial civilisations was just beginning to leave the confines of science fiction literature and filter through to academic circles, encouraged by the first successes of the space age. The considerable effort and enthusiasm of pioneers like physicist Phil Morisson and astronomer Frank Drake were an essential driving force that soon bore fruit. Reasonably realistic projects were drawn up, techniques developed and conferences organised to discuss the best strategy to adopt in scanning the sky for signals from a potential sister civilisation.

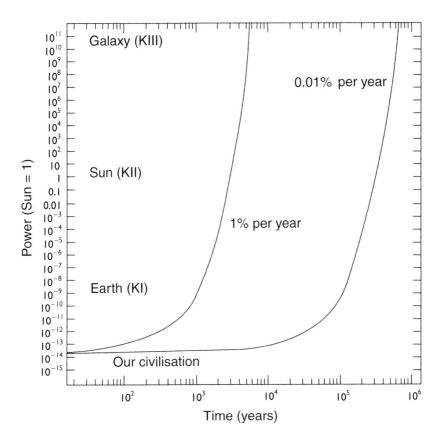

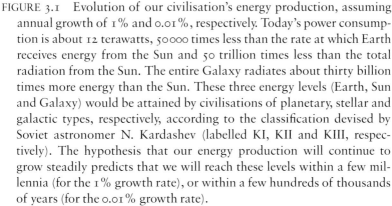

FIGURE 3.1 Evolution of our civilisation's energy production, assuming annual growth of 1% and 0.01%, respectively. Today's power consumption is about 12 terawatts, 50000 times less than the rate at which Earth receives energy from the Sun and 50 trillion times less than the total radiation from the Sun. The entire Galaxy radiates about thirty billion times more energy than the Sun. These three energy levels (Earth, Sun and Galaxy) would be attained by civilisations of planetary, stellar and galactic types, respectively, according to the classification devised by Soviet astronomer N. Kardashev (labelled KI, KII and KIII, respectively). The hypothesis that our energy production will continue to grow steadily predicts that we will reach these levels within a few millennia (for the 1% growth rate), or within a few hundreds of thousands of years (for the 0.01% growth rate).

Dyson believed it was 'much ado about nothing'. In his view, an extraterrestrial civilisation, even one extremely advanced in technological terms, would not necessarily devote much time and effort to sending messages across space. Unlike the shipwreck survivors who desperately hope their bottle will be found by their own kind, since they know that such beings exist, an extraterrestrial civilisation would be unlikely to send signals out to creatures that very likely do not exist at all.

On the other hand, such a civilisation might well, in Dyson's view, modify its own environment to such an extent that the changes would become detectable, even at astronomical distances. To justify this viewpoint, Dyson assumed that any project, no matter how colossal (or crazy!) it might be, could be achieved by this supercivilisation, provided it did not contravene the laws of nature. He also assumed that cost would be no obstacle and that the technology used would be understandable to twentieth century Earthlings. Obviously, none of these assumptions is particularly realistic and Dyson was fully aware of the fact. He simply wanted to show that his idea was possible, and not necessarily that it was plausible. He published a short article to this effect in the journal *Science* in 1960.

The greatest source of energy in a stellar system is the radiation from its star. However, most of this energy is wasted, lost forever in the vastness of interstellar space. Our Sun is at present radiating about four hundred trillion terawatts, whilst only one billionth or so is intercepted by the various planets in the Solar System. According to Dyson, an advanced civilisation could enclose its star by a spherical construction, in such a way as to intercept almost the whole of its radiated energy. What is more, there would be enough living space inside the sphere to satisfy a civilisation's population growth over millions of years.

Since no machine can use energy with 100% efficiency, part of the energy would be disposed of by radiating it into space in the form of heat. In the case of Earth, about one third of the solar energy received is immediately reflected into space. The rest, having been recycled in various ways in Earth's atmosphere, hydrosphere and biosphere, is re-emitted in the form of infrared radiation, corresponding to our planet's average temperature ($15\,°C$ or $288\,K$). Terrestrial infrared radiation (or that of the other planets) is extremely weak and could not be detected at astronomical distances. But the infrared radiation of a sphere billions of times larger than Earth would be easy to detect, even from several hundred light-years away, thus betraying the existence of a technological supercivilisation. Whence Dyson's advice for those who wish to locate extraterrestrials was to seek out infrared sources in the sky.

It remains to be shown how effective this approach might be, for there are other astronomical sources of infrared emission. Many sources have already been detected, notably by the American satellite IRAS. Launched in 1982, IRAS mapped the sky at wavelengths between 12 and 60 μm (millionths of a metre), thereby revealing the existence of myriad infrared sources both in our own Galaxy and beyond. Most of these sources can be attributed to interstellar dust, heated to a few hundred degrees by light emitted from neighbouring stars. How then would we distinguish a natural source of emission from the Dyson sphere of a possible extraterrestrial supercivilisation? One way might be to seek other signatures, emissions at other wavelengths, but the method is far from holding the promise originally suggested by Dyson.

However, independently of its usefulness in this context, the idea had an important impact on futuristic thought. Indeed, we could imagine that our own civilisation might one day be able to carry out such gigantic projects, eventually domesticating our life-giving star.

In fact, Dyson's idea was not completely original. In his book *Dreams of the Earth and Sky*, published in 1895, Konstantin Tsiolkovsky had already mentioned that a technologically advanced society could extend its activity into space in order to use a larger part of the Sun's radiant energy. However, one of the legendary figures of science fiction, Olaf Stapledon, came much closer still to Dyson's formulation. In his monumental work *Star Maker*, which came out in 1937, the hero recounts his voyage of initiation through the myriad inhabited worlds of our Galaxy and the Universe, in search of the supreme being, the Star Maker. The narrator tells us: 'The galactic community... resolved to pursue the adventure of life and of spirit... began to avail itself of the energies of its stars upon a scale hitherto unimagined.... Every solar system was now surrounded by a gauze of light traps, which focused the escaping solar energy for intelligent use.' In fact, Dyson admits that he was inspired by Stapledon's work (which is not well known to the general public). His idea of an enormous spherical enclosure around the Sun is nothing more than an elaborate extension of this short description from *Star Maker*.

Dismantling a planet

How would the Dyson sphere be built? In his laconic 1960 article, the author poses the crucial question: where can we find the materials needed to construct a spherical shell of area one billion times greater than the terrestrial surface area? The answer is: by dismantling Jupiter, the biggest planet in the Solar System. It has a mass about three hundred times greater than Earth. Once redistributed around the Sun, this matter could form a spherical shell about one metre thick with a radius equal to the radius of the terrestrial orbit. Inside this immense shell, covering a surface area 600 million times as big as Earth's, not only could humanity find the living space required for its demographic expansion, but it could also have access to almost the entire radiant energy of the Sun.

The idea of dismantling a whole planet is even more extravagant than the construction of the Dyson sphere itself. It constitutes the first vision of any human intervention on such a gigantic scale, far greater than the mere terraforming of a planet (which only modifies the planet's surface) or asteroidal mining. We may well wonder whether such a project could really be carried out, or indeed whether it would be desirable given the scientific importance that each body in the Solar System represents for planetologists. A hubris of these dimensions also runs the risk of catastrophic consequences for the rest of the Solar System.

It might be thought that Jupiter plays an important role in the gravitational equilibrium of the Solar System and that its disappearance would perturb the orbits of the other planets. After all it is twice as massive as all the other planets put together. However, this is not the case. The Sun, a thousand times more massive than Jupiter, dominates the dynamics of our system. Naturally, if this giant planet were to disappear, its satellites would have to seek new homes, unless they too were recycled into the construction of the great sphere. With regard to the question of scientific interest, it is reasonable to expect that all bodies in the Solar System will have delivered up their secrets a few millennia from now. They will then retain only a purely commercial interest, particularly if today's mentality still prevails at this distant epoch.

Without taking sides in the academic debate over the question of whether or not we should one day dismantle a planet, let us see how Dyson approaches the problem using only twentieth century technology. He does insist, however, that his calculations do not necessarily imply that this same method should ever be applied. He merely aims to demonstrate that a planet could be dismantled using only the kind of technology that can be conceived of today.

Dyson proposes accelerating the planet's rotation about its axis until centrifugal forces become greater than its internal cohesive forces. At this point the object will begin to break up, projecting material into space. Breaking point is attained when the planet's period of rotation drops to about one hour. This critical period should be compared with the twenty-four hour terrestrial rotation period, and also the present Jovian rotation period of ten hours, which is the shortest of any of the planets. The planet must therefore be forced to spin ten times faster and this is made all the more difficult by its considerable mass.

In order to accelerate the spin of a planet, Dyson suggests wrapping it around with an enormous metal grid into which a powerful electric current is injected. This creates an electromagnetic force which, applied in the right direction, would cause a slight acceleration in the planet's rotation. Slowly but surely the centrifugal force would increase, particularly at the equator. When the rotation period reached breaking point, the first fragments would begin to fly off the equatorial zones of the planet. As the enormous spinning top turned faster, more and more chunks would fly off into space to be captured by a gigantic system of magnetic nets. The engineers would thereupon collect up all the material they needed to build the Dyson sphere.

This is doubtless the greatest technological feat that could be conceived on the scale of the Solar System, but how much time and energy would be required to carry it to fruition? Dyson assumed a relatively mild magnetic field intensity induced by the metallic grid, with a value of around one hundred gauss (something like five hundred times the intensity of Earth's magnetic field). The planet would then have to be spun up for around a hundred thousand years before it reached one revolution per hour. The power required to supply the magnetic field would be of the order of one billion terawatts, almost a hundred million times greater than the current total production of our civilisation, and a hundred times more than the solar radiant energy received by all the planets. Enormous solar panels would have to be deployed, with an area hundreds of times greater than the planetary surface area, in order to trap enough energy for the project's requirements.

Once again the phenomenal magnitude of the quantities involved might lead us to view this as a simple aberration. Dyson implicitly recognises the fact, preferring 'to transpose the dreams of a frustrated engineer to an astronomical context', since he suggests that this project is more likely to be carried out by an extraterrestrial civilisation. He obviously fears that his wild dreams bring him closer to science fiction than to engineering, but he may be mistaken. We should not forget the lessons we have learnt from the past, nor the simple mathematics of exponential growth (which are valid up to the limits of a system). Only a thousand years ago, the whole of humanity made use of less energy than is released by a single nuclear reactor today. Moreover, a growth rate in energy production of one in a thousand per year, a rather modest estimate, leads to a thousand-billion-fold increase by the end of just thirty thousand years. It is not therefore completely unthinkable that a future civilisation may dispose of the energy required to construct the Dyson sphere.

Despite its appellation, the sphere would not be a rigid construction. No material could resist the enormous stresses that would occur in it. Indeed, no structure more than a few hundred thousand kilometres long can exist within a hundred and fifty million kilometres from the Sun (the radius of the terrestrial orbit). We must therefore imagine a vast number of tiny islands in space forming dozens of rings around the Sun, rather like the asteroid belt. These rings would be placed at varying distances from the Sun, to avoid intersection of their orbits, and would be inclined at different angles, so that they would cover the largest possible area of the imaginary sphere. Some of these islands could be used as colonies, whilst others would merely serve as collectors absorbing solar energy. This was indeed Dyson's original idea. However, some science fiction writers have misinterpreted the concept, and rigid Dyson shells have now become the usual variant in science fiction stories.

Impossible alchemy

In Dyson's original project, Jupiter, having a mass about three hundred times greater than Earth, was proposed for sacrifice. Jupiter is mainly composed of the light elements hydrogen and helium. A large part of the hydrogen is located inside the planet in metallic form. Such a substance is unknown on Earth, but predominates inside Jupiter because of the enormous pressures, billions of times greater than Earth's atmospheric pressure. The central nucleus is composed of ices and rock, with a mass equal to about twenty times Earth's mass.

It seems surprising that Dyson should not have taken Jupiter's rather unusual composition into account. The idea of the sphere is not in question here, only the suggestion that it could be constructed by dismantling the giant planet. Indeed, under normal conditions, hydrogen and helium are gases rather than solids. The planet's strong gravitational attraction prevents them from dispersing into space, and the huge pressures in the inner layers maintain hydrogen in the metallic state. Clearly, if Jupiter is dismantled, these two factors will be removed and the light gases will waste no time in dissipating out across the surrounding space. Only the heavy materials in the planet's rocky nucleus will remain (carbon, oxygen, silicon, etc.), and these comprise a mere 5% of Jupiter's mass, almost fifteen times the mass of Earth. Although this would suffice to construct the Dyson sphere, especially if extremely light structures were used, it would surely be the greatest waste of resources in history! Could we really justify dismantling the biggest planet in the Solar System, just to recuperate a few per cent of its total mass? On another scale, this is reminiscent of the way the ivory traders are currently massacring rhinos, just to sell their horns.

It was this wastefulness that encouraged Adrian Berry to propose another method, even more speculative than Dyson's approach. In his book *The Next Ten Thousand Years*, published in 1974, he suggests a way of exploiting almost the whole mass of Jupiter. He appeals to the process of controlled thermonuclear fusion, but on a scale not yet envisaged.

The Sun and most stars obtain their energy from fusion of hydrogen into helium within their cores, where temperatures can reach tens of millions of degrees. The process has not yet been achieved in a controlled way using terrestrial technology. However, we can reasonably expect that hydrogen fusion will be mastered during the coming century. In the cores of red giant stars, fusion of helium produces carbon and oxygen at temperatures of a few hundred million degrees. In the most massive red giants, fusion produces even heavier elements, such as silicon, calcium and iron, at temperatures of several billion degrees. All these elements are dispersed across the Galaxy by supernova explosions which mark the end of a massive star. Thanks to this stellar alchemy, heavy elements are synthesised from the lightest elements of all, hydrogen and helium, whilst the latter were produced in the hot Universe of the Big Bang (see Chapter 4).

Although the role played by stellar reactors in heavy element production has been understood for half a century now, no one has yet seriously envisaged an artificial process of this kind with a view to producing some specific material (at least, not on a macroscopic scale). Hydrogen fusion is already difficult enough to handle, and temperatures ten or a hundred times greater are needed for the other processes.

Berry slips lightly across the psychological barrier to suggest that one day thermonuclear super-reactors will be built, capable not only of hydrogen fusion to produce helium, but also of helium fusion to produce carbon, oxygen and then other elements. Hundreds of these super-reactors would be sent into the Jovian atmosphere to absorb its hydrogen content and transform it into heavier elements. The yield from such futuristic alchemy could then be torn from the planet's gravitational hold by means of powerful magnets (possibly superconductors), supplied with part of the energy released by those same nuclear reactions. This method would avoid the terrible waste involved in Dyson's method. More than half the mass of Jupiter could be recovered in a usable form.

Of course, Berry has no idea how these super-reactors could be built. Unlike Dyson, who extrapolates to gigantic scales a method which has already been understood, Berry is extrapolating a completely unknown method to that scale. But the real difficulty is that his method is quite simply unfeasible. Indeed, fusion of Jupiter's hydrogen would release as much energy as the Sun emits every hundred million years. If only a fraction of this energy were to dissipate into the Jovian atmosphere during the operation, the whole thing would volatilise into space.

It is quite clear that Jupiter will never be dismantled in the way suggested by Berry. It is nevertheless worth consideration because the basic idea, controlled thermonuclear transmutation of light elements into heavier elements, remains an attractive proposal. It is quite simply the age-old dream of the alchemists, who sought to transform some other material into gold with the help of the philosophers' stone. Furthermore, we may recall from Chapter 1 that Jules Verne's projectile, sent into orbit by the monstrous Columbiad, was a perfectly unsuitable means of transporting passengers. However, the basic idea behind it, sending people into space in a pressurised cabin rather than in a dirigible or on the wings of a bird, has remained.

Ringworld

As soon as it came out, the concept of the Dyson sphere created considerable excitement amongst science fiction writers. The most famous of the imaginary worlds invented on this basis was undoubtedly the one described by Larry Niven in his book *Ringworld*, published in 1970. Instead of a sphere, Niven imagines a simple rigid ring, 300 million kilometres in diameter (the same as that of Earth's orbit), around an imaginary star named EC-1572. The ring is a thousand times more massive than Earth, 2 million kilometres wide (five times the Earth–Moon distance), and has an area 6 million times greater than the surface area of our own planet.

According to the author, this immense area was the prime motivation for building the ringworld. Indeed, the civilisation which built it had previously been spread out over ten different stellar systems, whose wide separations (several light-years) had been plaguing them with communication problems. The ringworld gave them all the space they needed to live together. Using an accelerator (a kind of ultrahigh-speed train) around its outer edge, the inhabitants of this world were able to make a complete round trip in just three weeks.

In order to simulate Earthlike gravitational conditions on its inner wall, the ring spins around the star at a speed of 1200 km/s, forty times faster than the speed of Earth in its orbit around the Sun. This creates a year of about 9 days. Enormous panels measuring 4 million kilometres by 2 million kilometres are placed in orbit inside the system, periodically casting large regions of the surface into shadow. In this way, although the living area permanently faces the star, its inhabitants are

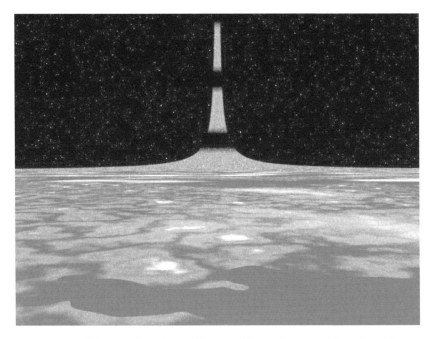

FIGURE 3.2   Ringworld, as it would appear from about 1000 km above the
ring. Bright and dark zones are alternating, the latter creating artificial
night in regions beneath the giant panels. (© James W. Williams.)

not condemned to live in eternal daylight. At the same time, these
panels absorb the star's radiation and transmit it to the ringworld,
thereby providing its main supply of energy. Despite the impressive size
of the system, the ringworld's own gravity is very weak. To prevent its
atmosphere from escaping into space, enormous walls are built along
the two edges of the ring, rather like mountains 1500 kilometres high.

The humanoid engineers of the supercivilisation built the ringworld
by gathering together all the solid materials in the stellar system EC-
1572 and transmuting them into a material of quite extraordinary
strength. The rigid ring could then resist the enormous stresses caused
by centrifugal forces. In addition, they took the precaution of collecting up all erratic objects in the stellar system, especially asteroids and
comets, to remove any risk of a collision later on, which might pierce
the ring and allow its atmosphere to escape.

It is hard to imagine the incredible sight that would meet the eyes of the ringworld's inhabitants when they chose to gaze upon the 'night' sky: a narrow band crossing the black sky from one side to the other, covered with light and dark stripes (the latter caused by the shadow of the panels). During the day, however, the illusion of a normal sky and a normal horizon would be perfect, except for those dwelling at the foot of the great mountain walls.

Despite Niven's obvious attempt at achieving an impression of realism, the ringworld could not exist. A construction of this kind suffers from a serious problem of instability: any small disturbance, like a meteor impact, would set the ringworld drifting into its sun. But with its almost limitless space, the ringworld has nevertheless stimulated the imagination of millions of science fiction readers.

Stellification of Jupiter and the Landau affair

Jupiter is the favourite candidate for most writers, serious or otherwise, when it comes to future development of the Solar System. This is probably because of its high mass. Dyson-style projects that aim to dismantle Jupiter and recover the materials making it up are certainly rather bold, but seem to violate none of the known laws of physics.

An even bolder project, and probably unrealistic, is to create a second sun in our system by transforming Jupiter into a star. The idea first surfaced in *2010: Odyssey Two*, Arthur C. Clarke's sequel to the famous *2001: A Space Odyssey*. In this second volume, Clarke tells how an extraterrestrial supercivilisation visits our Solar System and succeeds in triggering thermonuclear fusion reactions inside Jupiter. The aim of the operation is to endow the Jovian satellites with their own mini-Sun, thereby creating a mini Solar System with a dozen new worlds ripe for life. In other words, it is the first step in a vast planetary engineering project concerning the whole Jovian satellite system.

Although the idea is attractive, Clarke has some difficulty providing a convincing way of bringing about such a miracle. He no doubt remains consistent with himself and his third law, famous in the science fiction milieu: 'Any sufficiently advanced technology is indistinguishable from magic.' However, the irrefutable logic of this law should not be invoked to justify just any operation.

In fact, Jupiter is a failed star. Despite its great mass in comparison with the other planets, it nevertheless lacks the mass required to crush its innermost layers and raise their temperature to the point where hydrogen fusion can take place. It would need about eighty times its present mass to do that. The difference between a planet and a star is basically just a question of mass.

Clarke knows enough physics to be aware of this point. He therefore attempts to provide a plausible mechanism with the help of his character Vassili Orlov, the astrophysicist on board the Russian spacecraft Leonov. The vehicle is on a mission near Jupiter when the above-mentioned operation takes place, and the crew look on in stupefaction as the new star is born. To begin with, thousands of enormous machines gradually cover the surface of the planet, clearly robots of the von Neumann variety (see Chapter 2), multiplying exponentially from a single initial unit. Their role only becomes clear at the end of the operation. In Vassili's words: 'If a large percentage of Jupiter's hydrogen could be converted into much denser material – even neutron matter – that would drop down to the core . . . when the core became dense enough, Jupiter would collapse... and the temperature would rise high enough to start fusion.'

The description takes up the idea of a fusion super-reactor that can transform hydrogen into heavy elements. The observant reader will notice that it suffers from the same flaw as Adrian Berry's suggestion. In fact, fusion of only one tenth of the hydrogen on Jupiter into heavy elements would release as much energy as the Sun will radiate in the next 10 million years, and a fraction of it would certainly leak out. Not only would this energy be observed by the Leonov crew (something not mentioned at all in Clarke's account), but it would be released in a relatively short time and would vaporise the planet rather than cause it to collapse.

Apart from this detail, Clarke's suggestion is interesting. It was probably inspired by a forgotten but moving episode in the history of modern physics. The story goes back to the 1930s, the time of the Great Purge in Joseph Stalin's Soviet Union. Within a few years, over seven million people were arrested and almost three million executed. A large part of the Soviet intelligentsia was wiped out in these purges, including whole teams of scientific researchers. In such a climate, even a physicist as brilliant as Lev Davidovich Landau was exposed to the threat. Despite his young age, Landau was the best theoretical physicist of his day (and some specialists would say he was in the top ten of the twentieth century). Sensing the danger the young physicist thought he might save his skin if he attracted public attention by making some sensational discovery.

A few years previously, just after the discovery of the neutron, Landau had formulated the concept of the neutron star. This is a small and extremely dense stellar object, about a hundred thousand billion times denser than water, with a radius of order 10 kilometres and surface gravity billions of times greater than it is on Earth. Landau thought of using this idea to explain where the Sun and other ordinary stars got their energy. In fact, in the 1920s, British astronomer Arthur Eddington had already suggested that the prodigious quantities of energy released by our star might originate from nuclear reactions. However, the actual sequence of reactions involved was only brought to light in 1938 by the German physicist Hans Bethe, then living in the USA. For many physicists, the question remained open at the time. Landau thus imagined that there lay a small neutron star in the Sun's core. Attracted by its powerful gravitational field, the inner layers of the Sun fall inwards at speeds as great as one tenth of the speed of light. At the instant when they crash onto the hard surface of the neutron core, their kinetic energy is converted into heat which can hold up the weight of the Sun's outer layers before being radiated away from its surface. This process converts about 10% of the matter into energy, a proportion thirty times greater than that achieved by thermonuclear fusion of hydrogen. Such extraordinary efficiency in converting matter to energy would allow our star to shine for several billion years. In other words, the Sun's energy would be of gravitational origin, just like the energy released in hydroelectric installations, which comes from falling water.

Landau sent his article to the eminent Danish physicist Niels Bohr, asking him to submit it to the prestigious British journal *Nature*. He also arranged for the editors of the official Soviet newspaper *Izvestia* to print a commentary by Bohr. Although somewhat embarrassed by the request, Bohr immediately replied that the article seemed excellent and highly promising. A week later, Bohr's reply appeared as part of a flattering review of Landau's work in *Izvestia*. However, this spectacular device was not enough to save him. He was imprisoned a few months later, accused of spying for Nazi Germany, although the real reason was that he had criticised the Soviet government. Fortunately, Piotr Kapitsa, the greatest Soviet experimental physicist of his day, obtained Landau's release one year later by directly influencing Stalin. A few years previously, it was Kapitsa who had discovered the phenomenon of superfluidity in his laboratory. This is a state characterising certain fluids at temperatures close to absolute zero, where they manifest a total absence of viscosity (internal friction). Kapitsa requested Landau's liberation, arguing that the young physicist was the only Soviet physicist who might elucidate this mysterious and paradoxical property, thereby demonstrating Soviet scientific supremacy. Landau was duly freed and, a few years later, made a major contribution to the study of superfluidity, for which he earned the Nobel prize for physics in 1962.

After this lengthy historical detour, let us return to Clarke's idea for stellifying Jupiter. Is it possible on paper? The answer is negative. As shown by Robert Oppenheimer (father of the American atomic bomb) in 1938, the smallest possible mass for a neutron star is about one tenth the mass of the Sun. Less massive objects cannot exist in the neutron state because their gravity is insufficient to compress them to the required extent. If we could tear off a piece of neutron star, it would instantaneously explode due to the mutual repulsion of its constituent particles, which is exceptionally strong at such high densities. It is thus impossible to imagine transforming Jupiter into a neutron star. Its mass is only one hundredth of the minimal required mass.

In fact, the mere idea of increasing the density inside a star involves a further problem that could not have been conceived of, even by the gifted Landau. A star of this kind could never resemble the Sun or any other normal star. It would be more like the red giants, bodies with dense cores and extraordinarily inflated and rarefied envelopes. The structure of these spectacular objects (which are in fact aging stars) was investigated for the first time at the end of the 1930s by Russian physicist George Gamow and his Estonian colleague Ernst Öpik. Even today the exact mechanism leading to their formation is not understood in all its details. We do know, however, that in several billion years the Sun will become a red giant, hundreds of times brighter than it is at present.

Clarke was not the only one who tried to stellify Jupiter. More sophisticated projects have also been conceived, but none of them seem particularly convincing today. We may therefore hope that Jupiter will long remain in its present state, despite the threat of the would-be star makers.

Recounting the end of the world

The futuristic visions we have described up to now are resolutely optimistic. They assume that our species will manage not only to survive for several millennia to come, but also to extend its empire into space, increasing technological capabilities almost without limit. This is the position adopted in a great number of science fiction stories, especially during its golden age, up until the 1950s. These visions may seem naive today, but they cannot be excluded from our panorama of possible futures.

Other futures can be envisaged. Zero growth rates are rarely encountered in science fiction literature, probably because they leave little room for novelty, dreams and action. However, the present economic, ecological and other crises affecting our planet would suggest that it may be the only way we can survive, at least for a certain time, giving Earth respite to heal its wounds.

Once again, Arthur C. Clarke is one of the few science fiction writers to have treated this scenario with any degree of originality. In his book *The City and the Stars*, he describes the eternal city Diaspar, the ultimate and unique megalopolis remaining on Earth. Protected by a gigantic dome, this supreme achievement of Earthling technology has survived unaltered and intact for half a billion years into the future on an Earth almost completely reduced to desert. Lack of interest about the outside world stems from a traumatic encounter. It happened long ago as Earthlings were reaching out to conquer the stars and had suffered a terrible defeat when confronted by the overwhelming power of extraterrestrials. Stirred by curiosity, the young Alvin leaves Diaspar and discovers another city, Lys, much less advanced technologically but highly developed on the spiritual level. The two cities do have something in common, however. They have both long since ceased to evolve or to display any tendency for expansion, typical features of every living species. In an attempt to show the absurdity of eternal stagnation, Clarke sends Alvin and his companion from Lys on a trip to the stars, a first step towards regenerating the hibernating civilisation.

End-of-the-world scenarios are far more common. This is one of the richest and most thoroughly investigated themes in imaginative literature. But it is also among the hardest to treat. As so rightly pointed out by Jacques Van Herp in his fascinating *Panorama de la science-fiction (Panorama of Science Fiction)*: 'No theme is more revealing than the idea an author makes of mankind and its place in the Universe.' In these scenarios, human civilisation eventually disappears, wiped out by some natural disaster or self-destruction. Sometimes a few members of the human species may survive to try and build up a new civilisation. In other cases, the human race is entirely destroyed, bequeathing Earth to some different species. In yet other accounts, the whole planet, together even with the Sun and the entire Solar System, are completely annihilated, so that civilisation or whatever remains of it is forced to emigrate to the stars.

Such a pessimistic vision of mankind's future, reflected so well by Paul Valéry's remark 'we other civilisations, we now know that we are mortal', is relatively recent in the history of ideas. Scientific and technical progress in the nineteenth century had succeeded in creating a generally optimistic attitude towards the capabilities of the human species to control both nature and its own fate. It was quite inconceivable that humanity could one day disappear for good. But even at the beginning of the twentieth century, many works appeared announcing the apocalypse. This was before the Great War had broken out, and before the appearance of weapons capable of large-scale destruction. According to Van Herp, two factors had created this shift in viewpoint.

The first was the discovery of long lost civilisations. In the middle of the nineteenth century, Sumerian and Akkadian cities had been found in Mesopotamia, and in the 1870s, the excavations of Heinrich Schliemann had brought the Mycenaean civilisation to the public eye at the site of Troy in Asia Minor. Then the Cretan civilisation was revealed through the discovery of the Phaistos stone in 1908. All these finds suggested relatively developed and refined societies, sometimes superior to the civilisations which succeeded them. The idea that civilisation does not necessarily move forward in a continuous way, but rather vacillates between low and high achievement, was not a new one. The examples of Greece and Rome were well known. But the revelation that a civilisation could disappear without trace was something else.

The second significant factor that marked the beginning of the twentieth century was the almost total destruction of Saint-Pierre in Martinique, Messina in Italy and San Francisco in the USA. Within a single year, 1902, the three cities were devastated, the first by volcanic eruption, and the second and third by earthquakes. This had a profound effect on public opinion. It showed that, despite technological progress, Nature was still the more powerful. Mankind's fate lay in the hands of blind forces that could put an end to our existence on Earth at any moment. Many contemporary writers were to describe the disappearance of human civilisation as a result of volcanic explosions, cataclysm or other cosmic disaster. Shortly afterwards, the secret of the atom was unlocked, and the way was open to fabrication of weapons that could cause large-scale destruction. This only strengthened the general pessimism concerning the future of humanity.

Among those writers who have tried to imagine the long-term future of our species, Olaf Stapledon deserves to be singled out for his epic work *Last and First Men*, published in 1930. He considers the future of the human race on a quite unprecedented time scale, even in the context of science fiction.

Clearly influenced by the great German historian Oswald Spengler, Stapledon describes humanity's future as a succession of civilisations rising to their climax and then falling into decline. He begins with our own civilisation, the 'first humans', destined to end in a few thousand years following massive use of the atomic bomb, not yet invented in 1930! This leads to almost total extermination. A mere handful of men and women survive and humanity is swept into a barbaric period lasting ten million years. Then appears the second human species, which enters into a catastrophic war with the Martians (the similarity with Wells' *War of the Worlds* being striking). The new species is also swallowed up by an age of darkness, this one destined to last thirty million years.

In this way civilisations come and go on Earth for around two hundred million years. The last Earthlings are those of the fifth species. The Moon threatens to spiral in on Earth and they are forced to emigrate to Venus, after suitably modifying its hostile climate. This is the first mention of terraforming in the literature, as discussed in Chapter 1. It proves difficult to grow acclimatised to life on Venus and over the following seven hundred million years, three further species succeed one another on our sister planet, bearing no physical resemblance to ourselves. The eighth species, observing that the Sun is soon to blaze up in a collision with an interstellar cloud, employs bioengineering techniques (although Stapledon does not actually use the term) to create the ninth species of humans. Their fate is to confront the exceptionally hostile environment prevailing on Neptune, for which they were specifically designed. Transplanted to this distant planet, the 'human' civilisation then manages to survive a further billion years, once again evolving through alternating periods of greatness and decadence. Finally, it falls upon the eighteenth race to end the saga of human existence on Neptune, when they discover that the Sun has been destabilised by some unknown process and will explode within only a few thousand years, too short notice to undertake any form of rescue operation.

Stapledon was a philosophy teacher and knew nothing about science fiction. He wrote the epic *First and Last Men* as an essay on the future evolution of the human race in a cosmic context. Without a doubt, no one has better achieved, or will ever improve upon, the objective Stapledon set himself. As emphasised by Jacques Van Herp, 'the main features of this evolution are borrowed from mythology and theosophical cosmogony…, but Stapledon's particular contribution is this basic pessimism which decrees that, each time a species rises, it must turn to dust in war or self-destruction.'

Independently of these pessimistic or optimistic visions of our future, it is clear that cosmic catastrophes are a real possibility, threatening to wipe out all life on Earth. Even if the probability of such disasters is small, it is nevertheless not absolutely zero.

Celestial danger

The Apocalypse according to Saint John, which is the last book of the New Testament, is unique among eschatological writings (from the Greek *eschatos*, meaning last). This book was written around 95 AD and relates various events supposed to occur at the end of time, warning signs of the Last Judgement. One of the most powerful images is that of the seven angels blowing on seven trumpets, each one foretelling some particular disaster. At the sound of the second trumpet we are told: '. . . and as it were a great mountain burning with fire was cast into the sea: and the third part of the sea became blood.'

Asterix the Gaul and his brave friends were clearly not the only ones who felt the heavens might fall upon their heads. Shooting stars may well have been at the origin of this fear of celestial disaster. In Chapter 1 we noted that the scientific community was firmly against the idea of rocks falling from the sky until the end of the eighteenth century. But even after the idea was accepted, these falling rocks were never linked with such an apocalyptic disaster, at least, not until the twentieth century.

On 30 June 1908, a ball of fire shot across the Tunguska valley in central Siberia. A few seconds later, there was an enormous explosion which literally flattened 2000 square kilometres of forest (fortunately uninhabited). Millions of tonnes of dust were blown up into the stratosphere by the explosion. The dust scattered the Sun's light to such an extent that, ten thousand kilometres away in London, people were able to read their newspapers in the middle of the night. The Tsar's government, being so far away from this lost region of the immense empire, showed no interest in local accounts of this tremendous event. The first scientific mission arrived in the area in 1927 and found no trace of life. Around the point of impact, a region as big as Paris had been completely devastated, as if the ground had been ploughed over by thousands of bulldozers. Outside this zone, thousands of scorched trees lay flat in the aftermath of an enormous fire. Even further out, still more trees had been torn up or flattened to the ground by the shock wave from the phenomenal explosion.

Among the various hypotheses put forward to explain the Tunguska phenomenon, only the explosion of a projectile in the atmosphere seems plausible today. According to estimates, the object was a stony asteroid some thirty metres across and weighing several tens of thousands of tonnes. Entering the atmosphere at about 20 km/s (roughly 70000 kilometres per hour), the projectile was subjected to a tremendous pressure on its front face, causing it to explode before reaching the ground. This would explain why there was no crater within the devastated region. Had the object been composed of heavy metals, like an iron asteroid, it would have reached the ground, blasting out a crater more than a kilometre across and a hundred metres deep. A region as large as several thousand square kilometres would have been affected by the resulting earthquake and fallout of material torn up by the projectile. In all likelihood, it was just such an event (impact of an iron asteroid about thirty metres across) that made the famous Meteor Crater in the Arizona desert about fifty thousand years ago.

About a hundred craters of meteoritic origins and with diameters greater than one kilometre are known today over the surface of our planet. They clearly demonstrate that the threat of disaster from the skies is not an empty hypothesis. Had the Tunguska projectile arrived three hours late for its meeting with Earth, it would have wiped Moscow off the map.

Meteorite impacts on our planet are classified into four levels according to size. At the lowest level are objects which rarely reach the ground and cause negligible damage. These measure less than ten metres across and generally go unnoticed since they explode or vaporise in the atmosphere at high altitudes. The second level concerns objects measuring several tens of metres in diameter. They have kinetic energy equivalent to a nuclear bomb of a few tens of megatonnes, comparable with the most powerful nuclear weapons ever constructed (several thousand times more powerful than the 20 kilotonne Hiroshima bomb). The Tunguska and Meteor Crater events correspond to objects of this type. According to present estimates, we can expect one projectile with a diameter greater than thirty metres to collide with Earth about once every century. Given that about two thirds of Earth's surface is covered with water and less than one tenth of the land area is inhabited, victims would only be expected about once every few millennia. Very probably, no cosmic disaster of this level has ever occurred since the beginnings of civilisation on Earth.

At the top end of the second level are objects several hundred metres across. They have kinetic energy equivalent to several thousand megatonnes and could annihilate a whole continent. If one of these fell into the ocean, it would give rise to a gigantic tidal wave, initially a few metres high but reaching several tens of metres at the coast. Such an event is unlikely to have occurred since the appearance of Cro-Magnon man about fifty thousand years ago.

At the limit between second and third levels are objects measuring around a kilometre across. They have kinetic energy equivalent to about a hundred thousand megatonnes. From this point upwards, consequences of an impact affect the whole planet, rather than being limited to a region or a continent. Huge quantities of dust and ash (from forest fires) would be thrown into the upper atmosphere, blocking out the Sun's light for several months. Cast in darkness, plants could no longer photosynthesise adequately, thereby jeopardising the first link in the food chain. Moreover, the sudden fall in temperature would destroy most of the world's crops and lead to a generalised famine. The whole structure of human society (health, politics, economics) would be put to the test by this catastrophe and few countries would be able to cope. Considering the estimated frequency of this type of collision, it is unlikely that any object of this size has collided with Earth since the appearance of Neanderthal man about a hundred thousand years ago. This global disaster is similar to a nuclear winter. The term refers to everything that would happen if the world's nuclear arsenals (in particular, those of the USA and ex-USSR) were simultaneously exploded. The consequences of nuclear winter were assessed for the first time at the beginning of the 1980s, in terms of studies relating to meteorite impacts! The results considerably aroused public feeling against the military antagonism indulged in by the two superpowers at the time. Fortunately, the threat of nuclear holocaust has greatly diminished since then.

The fourth level on the scale of meteorite impacts is associated with objects in the neighbourhood of ten kilometres across. A mountain the size of Mount Everest, after crossing the atmosphere in two or three seconds, would make a crater at least a hundred kilometres in diameter and more than a kilometre deep, releasing an energy equivalent to several hundred million megatonnes. Heated to over a 1000 °C, billions of tonnes of glowing material would be projected into the atmosphere before falling across the entire planet, setting fire to whole continents. The huge quantities of dust, ash and soot thrown into the stratosphere would block out solar light for years. The surface of the planet would be plunged into total darkness and its mean temperature would fall by about twenty degrees. Most of the globe would be frozen, even in the middle of summer. The consequences for the whole food chain would be even more serious than in the case of a third level impact. Most terrestrial flora and fauna would be driven to extinction. After a few years, the Sun's light would begin to penetrate the atmosphere once again, heating the surface of the planet. Normally, Earth radiates away its heat in the infrared region of the electromagnetic spectrum, but this type of radiation would be efficiently absorbed and re-emitted back down to the ground by vast quantities of carbon dioxide propelled into the atmosphere by the impact. A monstrous greenhouse effect would reign for several thousand years, creating a much warmer climate than we are used to today.

According to present estimates, this type of event could happen about once every 100 million years. The disappearance of the dinosaurs has been attributed to an apocalypse of this magnitude occurring 65 million years ago. Palaeontologists have long known that about half the world's vertebrates and marine species suddenly disappeared around this time, at the boundary between the Cretaceous (K) and Tertiary (T) periods. At the end of the 1970s, a team from the University of Berkeley, led by physicist and Nobel prizewinner Luis Alvarez and his son Walter, showed that sediments belonging to the K–T boundary are extremely rich in iridium; in fact, by at least a factor of one thousand compared with adjacent sediments. Now iridium is a rare element in the terrestrial crust. The most plausible explanation for this discovery is an asteroid impact. The asteroid's iridium content would have been dispersed across the surface of our planet in a relatively short period of time. The total quantity of iridium in sediments at the K–T boundary, over the whole surface of the globe, has been evaluated at around half a million tonnes. Calculating from the mean iridium content of asteroids, this implies an object with a mass of a thousand billion tonnes and a diameter of about ten kilometres.

Although highly controversial at the beginning of the 1980s, this hypothesis is now widely accepted amongst scientists. The Chixculub crater, believed to be the trace of the impact, was discovered in 1991. This crater is 200 kilometres in diameter and is located to the north of the Yucatán peninsular in Mexico. Its age has been estimated at around 64.98 million years, with a margin of error of a hundred thousand years. The crater escaped detection for a long time, being submerged under water for the main part and covered over with large quantities of successive deposits. It is at present the second largest crater known on Earth, after the Vredefort crater in South Africa, which was discovered in 1993 and dates from two billion years ago.

There does exist a fifth level of impact, although it is of purely academic interest. It refers to collisions with objects more than 200 kilometres across. There is no such object amongst the nearby asteroids.

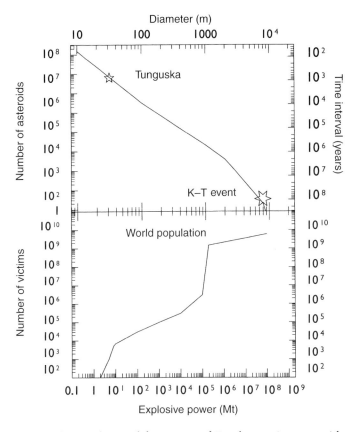

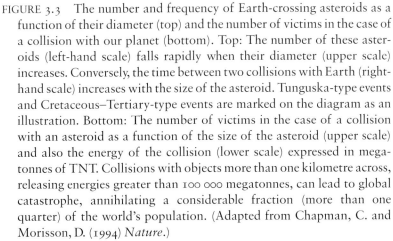

FIGURE 3.3   The number and frequency of Earth-crossing asteroids as a function of their diameter (top) and the number of victims in the case of a collision with our planet (bottom). Top: The number of these asteroids (left-hand scale) falls rapidly when their diameter (upper scale) increases. Conversely, the time between two collisions with Earth (right-hand scale) increases with the size of the asteroid. Tunguska-type events and Cretaceous–Tertiary-type events are marked on the diagram as an illustration. Bottom: The number of victims in the case of a collision with an asteroid as a function of the size of the asteroid (upper scale) and also the energy of the collision (lower scale) expressed in megatonnes of TNT. Collisions with objects more than one kilometre across, releasing energies greater than 100 000 megatonnes, can lead to global catastrophe, annihilating a considerable fraction (more than one quarter) of the world's population. (Adapted from Chapman, C. and Morisson, D. (1994) *Nature*.)

As we saw in the first chapter, the largest near-Earth asteroid, 1036 Ganymede, does not exceed 40 kilometres in diameter. However, several objects of this size do gravitate around the Sun in the distant asteroid belt lying between the orbits of Mars and Jupiter. Collision with a body of these dimensions would produce a crater about 1500 kilometres across (with an area three times as large as France) and more than 3 kilometres deep. A huge quantity of vaporised rock would be blasted into the atmosphere. The planet would be enveloped in a thick suspension of molten magma at 2000 °C. The heat radiated down to ground level by this blazing shroud would evaporate all the water in the oceans after a few years. During this time the cooling magma would eventually condense out, depositing a layer of rock 300 metres thick over the surface of the globe. An extremely thick fog, hundreds of times more dense than our present atmosphere and almost exclusively composed of extremely hot water vapour, would gradually cover the whole planet. After several thousand years, the vapour would in its turn cool and condense, filling the oceans once more.

Unlike the K–T-type events, no life form could survive the apocalypse ensuing such a high-energy impact, even on a microscopic scale. This kind of collision almost certainly occurred at the very beginning of Earth's history, four billion years ago, but erosion has since erased all trace. Several enormous craters on the Moon, such as Orientale Basin or Mare Imbrium, are witness to the violence of certain titanic collisions that marked the genesis of the Solar System. Fortunately, there are far fewer objects of this kind in circulation today. Indeed, Earth is very unlikely to encounter a body of this size for billions of years to come.

The sword of Damocles

Damocles was a courtier in Syracuse at the court of the tyrant Denys. He believed that his master's life was one of pure happiness. In order to prove that this was not the case, Denys invited him to share his privileges for a few days and Damocles immediately accepted. Seated upon the throne, he raised his eyes to the ceiling and saw a huge sword hanging there by an invisible thread and pointing towards his head. Denys wanted him to understand that a tyrant's throne is never safe from danger.

Is our civilisation safe from unpredictable cosmic threat? There is a tiny but nonzero probability that a sizeable asteroid will strike our planet and cause global destruction. Considering the great number of casualties in such a situation, the chances of any particular individual perishing in this type of catastrophe are not negligible. American astronomers Clark Chapman and David Morisson tried to evaluate the risks in an article which appeared in *Nature* in 1994. The impact of an asteroid measuring more than one kilometre across can be expected to occur on average once every 200000 years. This implies a probability of about 60 in 200000, or roughly 1 in 3000, that it will happen within the 60-year life span of a human being. Assuming that a quarter of humanity would perish in the event (according to the very definition of a global catastrophe proposed by Chapman and Morisson), we discover that each of us has one chance in 12000 of dying in this way. This is much lower than the risk of dying in a car accident (about 1 in 100), but much higher than the chances of dying in a plane crash! The latter probability is estimated, in a rather approximate manner, as follows. About 100 million human beings die each year (the total population of 6 billion, divided by the average lifespan of a human being, which is about 60 years). There are at least a hundred deaths by aircraft accidents (several dozen small accidents and the occasional crash of a large passenger aircraft). The probability of dying in this type of accident is thus slightly greater than one in one million. Naturally, this is only an average value. The risk is considerably greater for pilots and almost zero for those who never travel by plane.

It should be emphasised that the low frequency of these cosmic catastrophes does not imply that the next one will occur 200000 years from now. It might just as well happen in 500000 years, or in 20 years. The collision between comet Shoemaker–Levy and Jupiter in 1994 clearly demonstrated that this type of encounter, releasing almost a million megatonnes of energy, has been a regular feature in the history of our Solar System.

How can we find out if one of these bodies threatens to collide with our planet? How could we parry the blow from this cosmic sword of Damocles? Over the past few years, the scientific community has seriously turned its attention to these questions. Although a satisfactory answer to the first question will be forthcoming within the next twenty years or so, the second poses a far greater challenge.

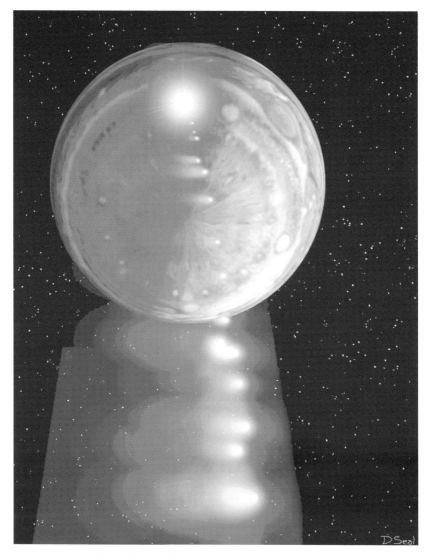

FIGURE 3.4 The collision of comet Shoemaker–Levy with Jupiter in
1994. The comet split in more than a dozen fragments, shown on their
collision course with the giant planet. The probability of Earth suffering such a collision is small, but not zero. (Image created by David Seal,
NASA/JPL.)

It is no easy matter to detect these space vagabonds. The smaller the object, the less solar light it reflects and the more difficult it is to spot. About a hundred nearby asteroids of diameter greater than one kilometre have currently been catalogued. However, their estimated number is more like 2000. At the present rate of discovery, several centuries would be needed to produce a complete list. In 1992, NASA studied a project aiming to identify systematically all nearby objects which could possibly cause global disaster on Earth. The scheme was called Spaceguard, named after the equivalent project in Arthur C. Clarke's famous novel *Rendezvous with Rama*. Spaceguard would comprise six telescopes of diameter between 2 and 3 metres, located uniformly across the globe. Equipped with state-of-the-art detectors, the system would scrutinise the whole celestial sphere in a permanent way. The images would be analysed for rapidly moving objects, since bodies close to Earth are characterised by high apparent velocity. The cost for setting up the Spaceguard mission would be about 50 million dollars, and it would cost a further 10 million dollars to operate per year. Around 20 years would be required to detect all near-Earth asteroids measuring more than one kilometre across.

Future civilisations will no doubt find a way of protecting themselves against asteroid impacts on Earth. However, no effective means of defence is known today. In the case of a relatively small object, we could only organise an evacuation of the region surrounding the point of impact. Unfortunately, the exact location of this point would only be known with any accuracy a few days before impact. In the case of massive objects the time factor is of capital importance. An effective parry could not be envisaged in under a few months, or even years. In science fiction, there is a widespread idea that megatonne nuclear weapons could be used to explode the projectile. But this would serve no useful purpose. An object of this mass cannot be pulverised. At best, it would be shattered into several dozen pieces, each measuring hundreds of metres in diameter, and these would continue along their fateful trajectories. This is precisely what happened to comet Shoemaker–Levy in 1994. It had long been fragmented by Jupiter's gravitational forces and the various pieces ploughed into the surface in quick succession. The destruction resulting from a series of projectiles of these dimensions showering down on Earth would be even more disastrous than if the asteroid had struck intact.

The only realistic alternative available today would be to cause an explosion on one side of the asteroid, in the hope of deflecting it from its original trajectory (see Fig. 3.3). If the explosion could be arranged several years before the fateful date, even a tiny initial deviation would suffice to avoid collision. If we had several decades, even a conventional (chemical) explosion might be sufficient.

Many science fiction writers have been inspired by the theme of a cosmic body threatening to crash into our planet. In general the body is named after some destructive divinity in Scandinavian or Hindu mythology. Such is the case in Gregory Benford and William Rotsler's story *Shiva Descending*, published in 1980. Shiva the destroyer is the third divine being in the Hindu pantheon, after Brahma the creator and Vishnu the preserver. The most successful work in scientific terms is perhaps Arthur C. Clarke's *The Hammer of God*, published in 1993. The action takes place at the beginning of the twenty-second century, at which time mankind has already colonised the Moon and Mars. Looming up from the distant confines of the Solar System, somewhere beyond the orbit of Jupiter, the asteroid Kali is detected on a collision course with Earth. The object is named after Kali the goddess of death in Hindu mythology. Once the alert has been given, a spacecraft sets out to meet Kali several months later. The crew fix a mass propulsion device on its surface, with the intention of deflecting it from its course. However, the instrument has been sabotaged by a fanatical sect and proves faulty, so that the crew have no choice but to attempt to push Kali using their spacecraft engines. After many mishaps, they manage to prevent a head-on collision and Kali merely grazes Earth's atmosphere, causing a hundred thousand deaths, before cruising off into space.

The problems involved in the defence against asteroids are not only technical. This has been emphasised in an article by Andrea Carusi, who presides the committee concerned with small Solar System objects at the International Astronomical Union: 'The question of defense against projectiles from space is rather a delicate one and should be approached with a great deal of caution. It should not provide a pretext for developing military technology.' Indeed, the idea of deflecting these bodies by means of nuclear explosives could be exploited by certain countries as a pretext for continuing nuclear testing or developing new weapons. Moreover, any country that mas-

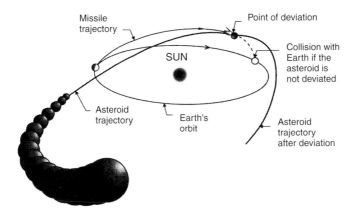

FIGURE 3.5 Defending Earth against an asteroid that threatens collision. The object is detected several months before its trajectory is due to intersect our own. A nuclear or chemical missile is sent out to explode near the asteroid and deflect it from its path. (Adapted from a NASA document, featuring in Sagan, C. (1995) *Pale Blue Dot.*)

tered techniques for deflecting such objects could presumably use those same techniques in the event of military conflict. In his book *Pale Blue Dot*, Carl Sagan illustrates the risk of developing such technologies by the story of the Camarina marshes.

In the sixth century BC, the inhabitants of Camarina, a little village in the south of Sicily, decide to drain the neighbouring marshland, which they believe to be responsible for a recent outbreak of the plague. Going against the oracle's advice to sit out the epidemic patiently, they go ahead with their plan, and to good effect. The epidemic soon dies out. However, in 552 BC, an army from the neighbouring town of Syracuse crosses the dried out marsh, which now presents no obstacle whatever, massacring all the people of Camarina and razing their village to the ground.

The story of the Camarina marshes shows how measures that have not been carefully thought through can sometimes lead to an even worse disaster than the one they were designed to avoid. Sagan advises the greatest caution in setting up an active defence against asteroids. Collaboration between scientists, military authorities and politicians together with a better informed public and improved international relations will be essential for the success of such a venture. If we follow these lines, the threat of cosmic disaster might just strengthen the sense of solidarity throughout the world.

An unstoppable catastrophe?

There is another type of cosmic catastrophe which may strike our planet in the very long term, namely a supernova explosion in the vicinity of the Solar System. Supernovas are amongst the most violent and energetic phenomena in the Universe and mark the end of a star's lifetime. During these explosions, a huge amount of energy is released. Over a relatively short time span, a few seconds, minutes or hours, a flash of X-rays and ultraviolet rays is emitted from the surface of the star, whose temperature reaches several hundred thousand degrees. The energy released in the flash is equivalent to the total energy radiated by the Sun over a period of several tens of millions of years. Ten times as much energy again is released during the following months in the form of gamma radiation, high-energy photons liberated in radioactive decay of atomic nuclei produced during the explosion. Finally, over a further period lasting anything between a few decades and a few centuries, ten times more energy again is emitted in the form of particles (protons, nuclei, electrons) accelerated by the explosion up to speeds close to the speed of light.

The consequences of such a tremendous wave of energy breaking over our planet clearly depend on the distance of the initial explosion. A supernova located at the distance of the nearest star, about 4 light-years away, would blast away the outer layers of Earth's atmosphere and heat the remainder to temperatures tens of degrees higher than their present mean temperature. Our planet would enter a period of violent storms, torrential rain, and extensive conflagrations, lasting for several years. All life forms on the terrestrial surface would be wiped out by intense bombardment from hordes of high-energy particles originating in the supernova. However, aquatic fauna and flora living at depths greater than a few metres would survive the catastrophe.

In his science fiction book *Inferno*, written jointly with his son Geoffrey in 1962, famous astrophysicist Fred Hoyle describes the terrestrial consequences of an even greater explosion. The nucleus of our Galaxy suddenly flares up. However, since it lies at 30 000 light-years from Earth, the effects on our planet are similar to those caused by a nearby supernova. Although the scenario is extremely unlikely, the physical consequences are described with exceptional skill.

The probability of a supernova explosion in the vicinity of Earth is actually negligible, even in the very long term. Indeed, our closest stellar neighbours will never be able to explode as supernovas, but instead will die peacefully like our own Sun. This is because they do not have high enough masses to do otherwise. The frequency of supernovas over the whole Galaxy stands at about two explosions per century. The frequency of such explosions occurring within a few light-years of Earth can be evaluated at less than one every 100 billion years. This shows that since the Solar System formed 4.5 billion years ago, there have probably been no supernovas in the terrestrial neighbourhood. Hence, we may view even the very-long-term future with some degree of serenity.

Obviously, the probability that a supernova will occur within a radius ten times greater, up to about fifty light-years away, is a thousand times greater. We would expect one every 100 million years or so. This is much longer than the present age of the human species, but it is not so long when compared with astronomical time scales. In fact, choosing 100 million years as the unit of time and calling it one 'cosmic year', we find that the Sun is already 45 cosmic years old and that it can expect to live another 65 cosmic years. It seems quite probable that, if our civilisation survives as long as Stapledon has envisaged, it will have to face such events on several occasions.

It is fortunate that the violence of the consequences of an explosion decreases as the square of its distance. The impact on Earth of an explosion ten times further away would be a hundred times weaker than the effects we described earlier. The greatest danger for life on our planet would then come from destruction of the ozone layer. This thin layer, only a few millimetres thick, is located at an altitude of 30 kilometres. It absorbs most of the Sun's ultraviolet radiation, which causes our skin to tan in the Sun as well as causing a great number of skin cancers. High-energy radiation from the supernova (X-rays, gamma rays or particles) would induce a complex chain of chemical reactions in the atmosphere, particularly those involving nitrogen-bearing molecules. These reactions could destroy 50% to 90% of atmospheric ozone over periods of anything between a few years and several centuries.

Present uncertainties, mainly related to the complexity of atmospheric chemistry, make it difficult to give an exact quantitative assessment. However, it seems clear that a nearby supernova explosion would have dramatic consequences for most life forms.

How could a future civilisation react to such a threat? The main problem stems from the fact that we cannot predict when a star will explode as a supernova. The X-ray flash and the first gamma rays will fall on the upper atmosphere at the same time as the light which announces the explosion, leaving insufficient time to react. It is true that most of the energy from exploding massive stars is carried away by neutrinos, and these subtle particles will arrive at least a few hours before the light. This is what happened on 23 February 1987 for supernova SN1987A in the Large Magellanic Cloud, a nearby galaxy. A few hours or days might just be enough to set up some kind of defensive shield. Enormous reflective panels would already have to be in orbit around Earth. These could then be positioned between ourselves and the supernova as soon as the neutrino signal was received. Several panels would be needed because the chances are that, under assault from the supernova radiation, they would soon melt down.

Can we predict the end of history?

In his authoritative work *Histoire de l'avenir, des prophètes à la prospective (A History of the Future, from Prophesy to Prospection)*, Georges Minois produces a fascinating panorama of the relationships various civilisations have held with the future. The first followers of the numerous Christian sects, as well as the Church Fathers up until the third century AD, believed that the end of the world was nigh. Many scholastic thinkers of the Middle Ages spent the greater part of their time working out the exact date at which the world would end, basing their study on readings from the Gospels and the Bible. The millenarian prophecy reached its apogee during the French Wars of Religion in the second half of the sixteenth century.

Today we cannot rule out the possibility that the human race will one day disappear, through either cosmic or anthropogenic catastrophe. In previous sections, we have seen that there is a nonzero probability of a nearby supernova or impact from a large asteroid. According to present estimates, this type of phenomenon occurs once every 100 million years or so. Recently, some scientists have used a new (and rather curious) type of statistics to predict the duration of our species.

These arguments are based on the principle of mediocrity: there is nothing extraordinary about us as intelligent observers of the Universe. The spatial version of this principle, usually associated with the name of Copernicus, has proven correct every time it has been confronted with observational evidence. Indeed, we live on a small planet, orbiting a very ordinary star among myriad others of the same kind, within an unexceptional galaxy among hundreds of billions of other galaxies in the observable Universe.

In 1993, American physicist Richard Gott applied the same principle to our temporal location within the Universe: our existence as intelligent observers at the end of the twentieth century should have nothing extraordinary about it. According to Gott, it would be exceptional if we were to arrive in the world either at the very beginning or at the very end of humanity. An intermediate temporal position, somewhere between these two extremes, is much more probable, and complies better with the principle of mediocrity. This temporal version of the principle would allow us to quantify humanity's life expectancy T_{future}, from knowledge of its present age T_{past}.

The probability that an observer should be born into the second or third quarter of the history of mankind (i.e., over a period including exactly half of the total history of mankind) is a priori 50% (see Fig. 3.6). An observer arriving right at the beginning of this interval would find that the first quarter of history ($= T_{past}$) had already happened, whilst three quarters ($= T_{future}$) remained. Conversely, the observer born at the very end of this interval, at the end of the third quarter, would find three quarters of history ($= T_{past}$) finished and one quarter ($= T_{future}$) to go. In the first case, we find $T_{future} = 3T_{past}$, and in the second case, $T_{future} = \frac{1}{3} T_{past}$. For other observers falling within the same interval, the future duration of mankind would take a value intermediate between $\frac{1}{3}$ and 3 times the past duration.

Hence, according to Gott, applying the principle of mediocrity to the temporal position of a human observer would allow us to deduce that the future duration of our species lies between $\frac{1}{3}$ and 3 times its past duration with a probability of 50%. In the same way, Gott finds that there is a 95% chance that the future duration will last between $\frac{1}{39}$ and 39 times our past duration (Fig. 3.6). If we take 100000 years as the

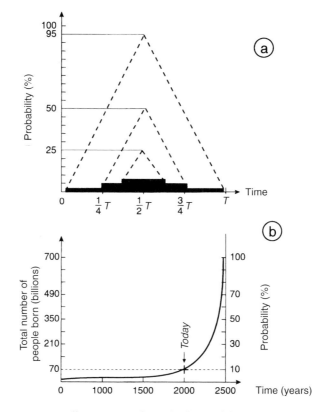

FIGURE 3.6. An illustration of methods used by R. Gott Jr (top) and J. Leslie (bottom) to assess the future duration of our civilisation, assuming finite total duration T. Top: According to Gott, the probability that we occupy a particular temporal position between 0 and T is the same for all points in the interval. This gives a probability of 50% that we lie in the range from $\frac{1}{4} T$ to $\frac{3}{4} T$, a probability of 95% that we fall between $\frac{1}{40}$ (0.025) T and $\frac{39}{40}$ (0.975) T, and so on. Since $T_{future} = T - T_{past}$, there is a 50% chance that T_{future} should lie between $\frac{1}{3}$ and 3 times T_{past}, a 95% chance that T_{future} should be between $\frac{1}{39}$ and 39 times T_{past}, and so on. Of course there is a 100% probability that T_{future} lies between 0 and an extremely large value (infinity). This argument teaches us nothing new. Bottom: Unlike Gott, J. Leslie used a clock that runs faster and faster as time goes by: the world population, in exponential growth. In his view, the probability of being in the first thousandth of the total number of humans to live on Earth is very small. Any given individual is more likely to fall within the great mass of

past duration of our history (since the appearance of *Homo sapiens*), there is a 95% chance that its future duration lies between 2500 years and 4 million years. Hence, concludes Gott, it is quite probable that the future duration of our species is limited to at most a few million years. Incidentally, biologists have found that most mammalian species to have lived on Earth have disappeared after 6 to 8 million years. On the other hand, if we adopt a value of 3 million years for the past duration (since the appearance of *Homo habilis*), the probable life expectancy of our species is found to lie somewhere between 800000 and 100 million years.

These arguments appeared in *Nature*. Readers' objections and Gott's answers to them were also published. Despite Gott's obvious confidence in his reasoning, I personally find the statistics extremely dubious. Pushing the arguments a little further, we find a probability of 98% that T_{future} lies between $1/100$ and 100 times T_{past}. Extrapolating to the extreme, there is a 100% chance that T_{future} take some value between 0 and infinity! We end up with a trivial result: (1) we are more likely to have a wide range of values than a smaller one for the future time span of history; (2) it is absolutely certain that the future time span will be somewhere between 0 and infinity. Clearly, this kind of argument teaches us nothing new. On more technical grounds, the argument suffers from errors in the application of statistical principles. In particular, it implicitly assumes that the longevity of our species is uniformly distributed in time, that is, that T can take any value with equal probability. Since we have no idea about the longevity of any intelligent species, this hypothesis cannot be sustained.

In his recent book *The End of the World*, Canadian philosopher John Leslie uses a similar argument, attributed to physicist Brandon Carter at the Meudon Observatory in Paris. According to the Leslie–Carter argument, we are more likely, at the end of the twentieth century, to be among the first tenth of the total number of intelligent

CAPTION TO FIGURE 3.6 (*cont.*)

observers arriving in the world during the last few centuries before the end of history. This would imply that the future history of our species is limited to a few centuries at the most. The problem with the argument is that, applied by future generations of observers, it would give them a few centuries to live, too, and this could go on for ever!

observers on Earth than to be among the first thousandth, or the first millionth. Given the exponential growth of the world population, the remaining tenths should come during the next few centuries. Consequently, Carter and Leslie find it very probable that the future duration of our species will not exceed a few centuries. This is much shorter than the time span calculated by Gott, because the Leslie–Carter clock, the total number of observers, runs ever faster due to the exponential growth of this number (see Fig. 3.6). In contrast, Gott's clock, recording normal time, always runs at the same rate. This change of clock leads to a major difference in the probabilities defined in the two cases. According to Leslie and Carter, it is highly likely that a given observer will fall within the great mass of observers born into the last few centuries before the end. In contrast, this extreme temporal position is considered as highly unlikely by Gott. For the same reason, the duration T_{future} calculated by Leslie is much shorter than the one found by Gott.

Leslie devotes a large part of his book to refuting various counter-arguments, often with some success. However, if his reasoning is valid, it must apply to observers in any epoch and not only to those living at the end of the twentieth century. By the same argument, a twenty-fifth-century observer (destined to disappear with the last generation of human beings, according to Leslie and Carter) would conclude that there still remained several generations of exponential growth for humanity before the end should occur. For this observer would con-sider it strange indeed if he or she should fall within the last thou-sandth of the total number of observers. Others living in the centuries and millennia to come would reach exactly the same conclusion, forever pushing back the end of humanity by a few more centuries. The fact that these observers are not yet born removes nothing from this counter-argument. If human history does end one day, it is clear that observers in that particular epoch will occupy an exceptional temporal position. By reserving the principle of mediocrity for our own period, we are depriving observers in the distant future from its benefits, thereby going against the very idea of the principle!

Leslie nevertheless admits that his argument is invalid if our future history is infinite. In this case, all observers will be located at the begin-ning of history. The past duration, always finite, will ever remain an infinitesimal fraction of what is to come. In the next chapter, we shall see that modern physics and cosmology allow an almost infinite playground for the future development of intelligence, although not necessarily of our own species.

The end of the Sun

Among the cosmic catastrophes, one in particular would seem to be inevitable. However, it is situated in a future so distant from us, on a time scale hundreds of times greater than any we have discussed up to now, that the word 'threat' loses all meaning. And yet, whenever it is mentioned, it incites a certain emotion: for we are speaking of the future and death of our own star, the source of all life on Earth, the Sun itself.

Born at the same moment as Earth around 4.5 billion years ago, the Sun has always led an exceptionally stable existence. With quite exemplary regularity, it has sent our planet almost a million terawatts of power, tens of thousands of times more than the global energy production of our civilisation today. This colossal energy intercepted by Earth's surface represents only a tiny fraction (roughly one billionth) of the total energy radiated into interstellar space by the Sun. However, the nuclear fuel reserves at the origin of this prodigious radiation are not infinite. At the present rate of consumption, in which around 7 billion tonnes of hydrogen are converted into helium every second, hydrogen reserves remaining in the Sun's centre will subsist for something like a further 6 billion years. A long death struggle will then begin, with dramatic consequences for the rest of the Solar System. In fact, Earth may become uninhabitable long before this date because of the gradual increase in the Sun's luminosity.

It is interesting to look at how the end of the Sun and Earth were conceived before the advent of modern astrophysics. At the beginning of the century, it was not understood where the Sun and other stars obtained their energy. No known energy source seemed able to supply the prodigious energy production of our star for billions of years. According to calculations by British physicist Lord Kelvin, if the Sun obtained its energy from gravitational contraction (exactly like our hydroelectric power stations), it could only shine for 30 million years at the most. This is far less than the age of our planet, estimated at the time to be around 2 billion years using radioactive dating techniques developed by New Zealand physicist Ernest Rutherford. When Einstein announced his special relativity theory in 1905, new solutions became possible through the postulate that mass and energy are equivalent. A small quantity of matter of mass M can be transformed into a staggering quantity of energy E, where $E = Mc^2$ and c is the speed of light. If the Sun drew its energy from conversion of matter, with 100% efficiency, it would shine for another 10000 billion years. This extraordinary life span, 1000 times longer than the present age of the Universe, is invoked in Olaf Stapledon's *First and Last Men*.

We know today that the Sun's life span is about a thousand times less than Stapledon had thought. Indeed, solar conversion of mass into energy has about 1% efficiency, typical of thermonuclear fusion of hydrogen into helium. Moreover, this concerns only the central tenth part of the Sun's total mass, the rest being too cold to burn (relatively speaking). These effects imply a life span of only 11 billion years for our star.

Even though the life expectancy of the Sun was not known at the beginning of the century, its eventual death had already been considered by Herbert George Wells in 1895 when he wrote his famous novel *The Time Machine*. For some, this story marks the beginning of science fiction itself. His time traveller passes through future centuries on board his machine, passively observing the evolution of our planet. Towards the end of his trip, he finds himself on an Earth undergoing its final death throes, all trace of civilisation having long since disappeared from its surface. Only the occasional primitive creature has managed to survive in the desolate landscape, under the pale glow of the dying Sun.

> So I travelled, stopping ever and again, in great strides of a thousand years or more, drawn on by the mystery of the earth's fate, watching with a strange fascination the Sun grow larger and duller in the westward sky, and the life of the old earth ebb away. At last, more than thirty million years hence, the huge red-hot dome of the sun had come to obscure nearly a tenth part of the darkling heavens ... The darkness grew apace; a cold wind began to blow in freshing gusts from the east, and the showering white flakes in the air increased in number. From the edge of the sea came a ripple and a whisper. Beyond these lifeless sounds the world was silent. Silent? It would be hard to convey the stillness of it. All the sounds of man, the bleating of sheep, the cries of

birds, the hum of insects, the stir that makes the background of our lives – all that was over. As the darkness thickened, the eddying flakes grew more abundant, and the cold of the air more intense. At last, one by one, swiftly, one after the other, the white peaks of the distant hills vanished into blackness. The breeze rose to a moaning wind. In another moment the pale stars alone were visible. All else was rayless obscurity. The sky was absolutely black.

Wells was probably aware of the Kelvin time scale, but his moving description of the Sun and Earth's ultimate future bears little resemblance to the picture provided by modern astrophysics. Indeed, using numerical computer simulations, astronomers now believe they have sufficient understanding of stellar evolution to be able to sketch with some accuracy the details of the Sun's future. They thus predict that its luminosity will increase slowly but steadily, by about 10% every billion years. After 3 billion years, solar luminosity will have risen by 30% or so and Earth will receive energy at the same rate as Venus today. The heating this will cause could lead to an uncontrollable greenhouse effect, equivalent to the events occurring on Venus itself in the past.

The progressive increase in the temperature of the globe will cause Earth's oceans to start evaporating, filling the atmosphere with a thicker and thicker fog. Since water vapour contributes to greenhouse effects, preventing ground heat from escaping into space, the atmospheric temperature will rise still further, thereby accelerating evaporation. It is not yet possible to describe exactly how these physical effects will operate. Vapour clouds play a double role of reflecting the Sun's light and absorbing ground heat, which makes it difficult to evaluate what will happen with any accuracy. It is clear, however, that the content of the oceans will eventually end up in the atmosphere, floating above a dead landscape, for all life forms will have long since disappeared.

Little by little the fog will dissipate. Under bombardment from solar radiation, water molecules will dissociate into their hydrogen and oxygen components. Being very light, hydrogen will escape into space, whilst highly reactive oxygen will start fires in the wastelands of dried-out flora, or else will be absorbed into rocks. In this way, after a few million years, and once the cloud cover has completely disappeared, Earth will display its new face. The ex-blue planet, cradle of life, will have metamorphosed into a completely sterile lunar-type landscape.

Considering the very long period of notice given for this disaster, we can reasonably hope that the human civilisation of this distant future will have had the time to think of a way of protecting itself. An obvious solution would be to set up enormous protective panels in the Lagrange point L_1 of the Earth–Sun system (see Chapter 1), each one measuring several thousand kilometres across. The opacity of these great shields could be adjusted to let through just the present amount of solar energy. The rest would be reflected or absorbed. Some of the excess energy could be put to use either on Earth or elsewhere. Similar measures would obviously have to be implemented with regard to the inhabitants of other sites in the Solar System, such as Mars, Venus or the Moon. Through such techniques, climatic conditions could be maintained at an acceptable level in each of these worlds for another 3 billion years or so. This takes us along to 6 billion years from now, 10.5 billion years after the birth of the Solar System. Civilisation must then come to terms with a genuine survival problem.

Chronicle of a death foretold

The hydrogen in the centre of the Sun will be exhausted and solar luminosity will be more than double its present value. A thin layer of hydrogen around the centre will then begin to burn, causing expansion of the star's envelope. This expansion will proceed at an imperceptible rate for about 700 million years. A slight orange tint will be the only sign that the Sun is indeed metamorphosing. But the evolution will steadily accelerate. The star will swell up quite disproportionately and its luminosity will rise to ten, a hundred and then a thousand times its present level. By now the Sun has turned red, its surface temperature having fallen to a mere 3000 °C. Towards the end of this period, and within only a few tens of millions of years, the radius of the star will extend from 40 million kilometres, the present radius of Mercury's orbit, to 150 million kilometres, taking its surface out as far as Earth.

Mercury will be devoured by the expanding red giant. It is not known what will happen to Venus, but Earth will in fact be spared due to a fortuitous outward shift of its orbit. This will take place without any human intervention. Indeed, during the whole of the expansion period, the outermost layers of the Sun will be projected into space by the high pressure in its interior. Their escape will be made easier by the fact that they originate far from the centre, where most of the Sun's

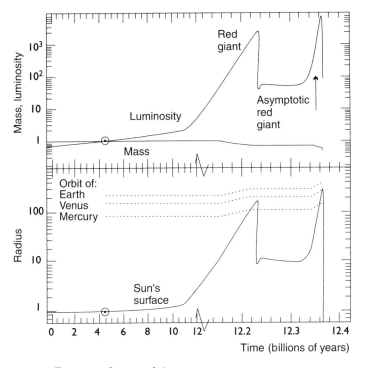

FIGURE 3.7   Future evolution of the Sun (top) and consequences for orbits
of Mercury, Venus and Earth (bottom). Top: Evolution of the Sun's
mass (expressed in units of its present mass $M_\odot = 2 \times 10^{27}$ tonnes) and
luminosity (expressed in units of its present luminosity $L_\odot = 2 \times 10^{26}$
watts). Once all core hydrogen has been consumed (6 billion years from
now), the Sun will become a red giant and its luminosity will increase by
a factor of 1000. Slightly later, helium will begin to burn in its core and
its luminosity will fall to forty times its present value. When all the
helium has been used up in this way, the Sun will begin to swell up once
more, becoming an asymptotic red giant. During this short period, it
will be even brighter than before. During its two giant phases, a large
part of its mass will be lost (in fact, almost half) through the stellar
wind. Its gravitational attraction for the bodies in the Solar System will
thereby be reduced. Bottom: The changing radius of the Sun (expressed
in units of its present value $R_\odot = 700\ 000$ kilometres) and evolution of
the orbital radii of Mercury, Venus and Earth (in the same units).
During its red giant expansion the Sun loses mass and consequently
relaxes its gravitational hold on the planets by about 40%. Less strongly
attracted, the planets gradually migrate outwards. The Sun will devour
Mercury, but Venus and Earth will escape.

mass is located, and will therefore feel its gravitational attraction to a much lesser degree. Ejected material will reach speeds of several hundred kilometres per second. In view of this haemorrhage, the Sun will lose matter at a rate of one terrestrial mass every millennium. By the end of this stage in its life, the Sun will have slimmed down to 70% of its original mass. Its gravitational hold on neighbouring objects, which is directly proportional to its mass, will likewise have been reduced by 30%. Being less strongly attracted by the Sun, the various bodies in the Solar System will migrate gently outwards. Venus will find itself where Earth is today, and is very likely to escape being swallowed up in the expanding red giant. However, it will begin to boil over, reaching a temperature of almost 3000 °C, as it grazes the surface of the red giant that will dominate nearly the whole of its sky. Meanwhile, Earth will be located only 60 million kilometres beyond the orbit of Venus. If an observer could survive the fiery furnace on its surface, at temperatures approaching 2000 °C, he or she would see a sight worthy of Dante's inferno. The Sun's disk would occupy more than three quarters of the sky.

This expansion will not be without consequence for the outer Solar System, although the giant planets themselves will be barely affected. As we saw in Chapter 1, three of Jupiter's satellites, Europa, Callisto and Ganymede, contain enormous amounts of water under their icy crust. It is also suspected that there may be water ice under the methane and nitrogen-rich atmosphere surrounding Titan, Saturn's largest satellite. The Sun's expansion will thus release vast quantities of liquid water over a period of several hundred million years. But afterwards, a large part of this huge reservoir will have evaporated and dissipated across space. The same fate awaits the water presently trapped in the smaller satellites and rings of the giant planets. The enormous reserves of water in Kuiper belt and Oort cloud comets will nevertheless remain intact.

Around the year 7 500 000 000, the whole of the Solar System will have become uninhabitable. Indeed, even at the distance of Pluto's orbit, forty times further out than Earth, the luminosity will be several times greater than its present level on Earth. But before considering the fate of civilisation in this distant epoch, let us follow through the final stages of the Sun's life.

After its long expansion period, and having threatened to swallow up everything in its path, the Sun will suddenly become well-behaved again, at least for a certain time. In its core, a helium cinder remains from the past combustion of hydrogen. When the core temperature rises to a hundred million degrees, the helium will start to burn, and this will extinguish combustion in the peripheral hydrogen layer. Now it was the latter which had caused the excessive expansion of the outer envelope. Like an ebbing tide, the star's envelope will begin to shrink, withdrawing from the burnt-out remains of Venus and finally stabilising at a radius of only 7 million kilometres, ten times its present extent. Alas, Mercury will have vaporised by then within the immense furnace, and will never be seen again.

For a further 100 million years, the half-red half-giant Sun will stay in this stage, shining with luminosity about forty times its present value. But its helium reserves will be exhausted much more quickly than the hydrogen it burnt before. The brief period of respite will soon reach an end. The Sun will then enter the final stage of its evolution. Combustion occurs in two layers at once: the helium shell around the core and the hydrogen shell even further out. This is the most unstable phase in its long history. Pushed by the spasms of the dying star, the outer envelope will once again begin to invade planetary horizons. During this relatively short period, lasting only half a million years, the greater part of the envelope will be expelled into space. For the first time, the Sun's core will be exposed, an incandescent but tiny sphere no greater than Earth. With a surface temperature of 100000 °C, it will radiate mainly in the ultraviolet. For several tens of millions of years, it will blow out a final kiss of death to the planets it had once offered life.

And so the Sun will die. Having a mass equal to only half its initial value, it will not succeed in triggering further nuclear reactions within its core. Gravitational contraction will no longer be able to create the compression needed to burn the remaining carbon. The stellar corpse, originally so hot, will gradually cool and it will become a white dwarf, dead for all eternity.

Moving Earth

Even though the end of the Sun will not come for an extremely long time, it is nevertheless quite unavoidable. Could our descendants survive, and if so, how?

A first reaction, in fact an innate reflex of every living creature, is to flee before danger. People may therefore consider evacuating all the worlds inhabited at the time (Earth, Mars, the Moon and possibly Venus and artificial colonies). The most obvious solution would be to lodge the whole population of the Solar System in artificial habitations, enormous space colonies, floating beyond the orbit of Pluto. In principle, asteroids would provide all the materials needed to build shelters for tens of billions of people. However, little is likely to remain of the asteroids by this time, for they will have long since been commandeered for other projects. Small satellites, or perhaps a small planet like Mercury, condemned anyway in the long term, could be dismantled. Clearly, the task would be made much easier if a Dyson sphere had already been built, comprising a large number of artificial planets that would gently migrate outwards from the Sun, using its energy.

This would provide a temporary solution but it would not save the cradle of our civilisation. Earth's surface would be scorched by the red giant and would become uninhabitable. For this reason, some writers have imagined moving the whole planet out of reach from the Sun's wrath. Motivations here are clearly rather sentimental, but it is nonetheless interesting to see how a supercivilisation might approach the problem. In the words of Archimedes: 'Give me something to hold on to and I will move the Earth.' Naturally, this great mathematician and engineer of antiquity wanted to bring out the technical potential of levers. However, these techniques are not applicable in the case of a heavenly body. In empty space, if we wish to displace an object, the only techniques available are those based on the principle of action and reaction.

It is easy to calculate the energy required to extract our planet from the Sun's gravitational attraction and push it towards the outer Solar System. It turns out to be equivalent to the energy radiated by the Sun during one year, far less than would be needed to dismantle Jupiter. The problem is more delicate, however. It involves transporting an inhabited planet together with the whole of its biosphere, quite a different matter from displacing an inanimate body. In addition, the planet is spinning on its axis once every 24 hours and this rhythm must not be disturbed during transportation. Any perturbation could cause a major climatic or ecological upset.

The operation must clearly be carried out very gently. This is quite feasible, since it will take the Sun 100 million years to swell up into a red giant, leaving ample time for such manoeuvres. The planet could be allowed to spiral out slowly towards the outer Solar System. Naturally, it will still orbit the Sun during the trip, but the year will become longer as time goes by. When its trajectory crosses the orbit of Jupiter, the terrestrial year will last ten of our current years. At the end of its voyage, slightly beyond the orbit of Pluto, the year will drag on for almost three centuries. Care must be taken to avoid collisions with other objects in the Solar System when Earth is moving across their trajectories. In particular, the whole asteroid belt will have to be cleared up beforehand.

This would indeed be the greatest spaceship ever envisaged (although in fact we have already been its passengers for millions of years). Concerning the means of propulsion, let us review the method devised by M. Taube, an engineer at the Institut Polytechnique de Zurich. The main difficulty is to preserve Earth's rotation about its own axis.

A system of 24 batteries of rockets would be installed around the equator. Each battery would be about 1600 kilometres from its nearest neighbours. There is a high probability that most of these batteries will be located right in the middle of the oceans, although it is very difficult to predict exactly where the continents will be in 6 billion years from now, due to drift of the tectonic plates, huge chunks of the terrestrial crust. Each battery would comprise a hundred rockets, each one with its exhaust nozzle reaching up thirty kilometres to stick out above the clouds, so that ejected gases could not disturb the atmosphere. (As we observed in Chapter 1, thirty kilometres is about the maximal height that could be envisaged for any construction on the terrestrial surface.) Each battery would be fired for one hour every day, at about 11 h 30 local time, so as to exert its thrust towards the Sun. The power released by each of these gigantic rockets, equivalent to a 1 megatonne bomb every second, would be used to heat up 100 tonnes of hydrogen per second, expelling it into space at a speed of 300 km/s. The resulting minuscule acceleration would barely be felt by Earth's inhabitants. The planet would thus reach its destination just beyond the orbit of Pluto after a few tens of thousands of years. The enormous quantity of hydrogen needed for the journey, almost one tenth the mass of Earth, would have to be borrowed from Jupiter or another giant planet.

This scheme appears to be the most detailed proposal yet made for a large-scale transportation of Earth (to my knowledge). It seems perfectly feasible for a technological supercivilisation. However, since we are dealing with a living planet rather than an inanimate body, it is hard to evaluate the consequences for the biosphere. Let us just bear in mind that this future civilisation is trying to save not only its own existence, but the very cradle of its origins.

Chronicle of an extended childhood

The finer part of mankind will in all likelihood not perish, they will migrate from sun to sun as they go out. And so there is no end to life, to intellect and the perfection of humanity. Its progress is everlasting.
Konstantin Tsiolkovsky, *Dreams of the Earth and Sky*

We have reviewed two different attitudes to the lethal threat of the Sun's future expansion when it becomes a red giant: hurried flight on light vessels or a farsighted wholesale evacuation of the entire planet from oncoming dangers. There is, however, a third reaction that might be adopted by a civilisation in the distant future, if it ever succeeds in reaching the level of a type II civilisation.

According to this classification, devised by Soviet astronomer Nikolai Kardashev, a type I civilisation is able to administer the entire energy and material resources of a planet. Our own civilisation today is not far from achieving this level (see Fig. 3.1), although our control over the situation leaves much to be desired! A type II civilisation is capable of controlling the entire energy and material resources of its stellar system, including the energy production of its own star. The Dyson sphere builders will clearly have reached this level. A type III civilisation on the other hand can manage the resources of a whole galaxy.

In principle a type II civilisation could domesticate its main energy supply, drawing upon it at will. Could such a civilisation go as far as extending the life of its star?

However fantastic it may seem, this idea has already come to light. In his book *Atoms of Silence*, Canadian–French astrophysicist Hubert Reeves sets out a solution for the well-being of his great great grandchildren. As we have already observed, the Sun's life will approach an end once all the hydrogen used as nuclear fuel in its core has been exhausted. But it will then have used up only about 10% of its total hydrogen reserves, those of the central regions, where temperatures are high enough for nuclear fusion to occur. Huge quantities of hydrogen will still remain intact around the solar core. Indeed, it is precisely the burning of peripheral hydrogen reserves that leads to its swelling up into a red giant. We might therefore imagine injecting fresh fuel into the core from outer regions, in order to resuscitate the failing central reactor. To achieve this, we would have to stir up the Sun's innards, rather as we stir our tea to mix in the milk and sugar.

To this end, H. Reeves has suggested that energy could be injected into the Sun to heat it up, thereby inducing vast convection currents. It is known that this kind of convective motion actually takes place within more massive and hotter stars than our own. Energy could be injected by means of a superpowerful laser, whose beam would penetrate right through to the innermost layers of our star. In this way, a further 10% or 20% of the Sun's mass could be made to burn, extending its life span by a factor of two or three, to reach 20 or 30 billion years.

There is an obvious difficulty with this scenario. How could energy be deposited inside the Sun without simultaneously heating the outer layers? The latter would cause them to dilate, thus creating a giant star before its day. But there is also a difficulty related to the sheer quantities involved. A quite colossal amount of energy would be required to heat up just a tenth of the Sun's mass to 10 million degrees (the temperature prevailing in the central regions of our star). In fact, as much energy would be needed as the Sun radiates over several million years. Since the energy must be injected fairly quickly to achieve the desired effect, it is quite clear that a truly prolific source must be found. The Sun itself would not fit the bill, its energy production being too slow. Explosive fusion of part of Jupiter's mass would suffice, but the consequences of such an explosion would be catastrophic for the rest of the Solar System.

Another astro-engineering idea has been put forward by Dave Criswell, ex-NASA engineer and research scientist in the American space industry. Criswell's suggestion for extending the Sun's life is to reduce the weight of its outer layers and hence benefit from a feature that stars have in common with humans: lighter beings live longer! In the case of human beings, this rule is of a statistical nature and of limited effect, since it refers to only a small percentage increase in life

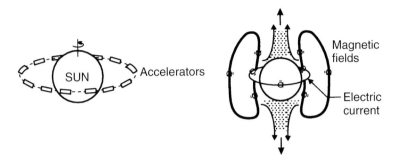

FIGURE 3.8   Lightening the Sun's load, as proposed by American engineer
    Dave Criswell. Left: A belt of powerful particle accelerators, supplied
    by solar energy, is set into orbit around our aging star. Right: The elec-
    tric current created by charged particles circling around the Sun induces
    a magnetic field. The latter exerts a force on particles at the solar
    surface, tearing them away and channelling them along the magnetic
    field lines to a storage area.

expectancy. In contrast, it is valid for all stars and the variation in life
expectancy for different masses is quite spectacular. A star with only
half the mass of the Sun will live 100 billion years, ten times longer
than our own star. This is because it burns its fuel ten times more
slowly and radiates ten times less.

Criswell is almost as bold as Dyson in his extrapolations. The huge
amounts of energy required for his project could only be supplied by
the Sun itself. The first stage would therefore be to deploy a set of solar
panels around the Sun, capable of intercepting a significant fraction of
its radiated energy. Construction materials could be extracted from
nearby Mercury, or else from further afield in the asteroid belt.

The second stage consists in wrapping the Sun around with a magnetic field that forces particles in the Sun's polar caps to fly out into
space. Such a configuration could be arranged by building a giant particle accelerator in orbit around the solar equator. In fact, a complete
belt of accelerators would be needed, placed several tens of thousands
of kilometres apart (Fig. 3.8). Accelerated by this equipment, an
intense beam of charged particles would orbit around the star, producing a powerful magnetic field with the appropriate features.

However, the magnetic field obtained in this way is not intense enough to tear particles in the solar surface from the star's gravitational hold. A third stage would thus involve heating the Sun's polar caps up to a few million degrees by means of a laser beam. The latter would be powered by the orbiting solar panels mentioned above. Heated gases would tend to disperse into space but in this case they would be channelled by the magnetic field. They could be stored for later use. By devoting one tenth of the Sun's radiated energy to this task, Criswell finds that we could thus extract a quantity of hydrogen equal to the mass of our own planet every 1000 years.

As this extraction progressed, its effects would gradually become observable. The luminosity of the slimmed down star would diminish and it would grow paler. Several hundred million years would be required to remove half the Sun's mass and thereby produce a star with life span in the area of 100 billion years, ten times its previous life expectancy. Of course, the new dwarf star would be ten times fainter than today's Sun, but then we did assume that the main concern of the future civilisation was self-preservation and not an increased energy consumption! And whenever this revitalised star came to the end of its days, the same process could once again reduce its mass by half to obtain yet another ten-fold extension. Needless to say, Earth would need to be pushed a little closer to the truncated star to ensure that it continued to receive the same amount of light. This ought to be a much easier task than slimming down the star.

Criswell's estimates are somewhat exaggerated. A star's life expectancy depends not only on its mass but also on the state of its core. Now the Sun has already transformed half the hydrogen in its central regions into helium. Criswell's half-Sun would therefore begin its new life middle-aged, and not a nursling in the cradle. Its new life expectancy would be a mere 50 billion years at most (and probably much less, owing to other more technical factors). But the scheme hides another problem. Astrophysicists believe today that when a star is small with correspondingly low luminosity, the inhabitable region around it (defined as the region in which temperatures are high enough to allow liquid water to exist on planetary surfaces) soon becomes very limited in size. On the other hand, when a planet is closer to its star, tidal forces are stronger. These tend to slow down its rotation about its own axis. In the end, the planet spins on its axis at the same rate as it moves about the star, so that one hemisphere faces the star all the time (just as happens to the Moon in its motion relative to Earth, shown in Fig. 1.1). There is then a large temperature difference between the two hemispheres which makes life there rather problematic. It turns out that stars of mass less than 70% the Sun's mass do not have continuously inhabitable zones around them. Hence it would appear that only 10 or 20 billion years' reprieve could be had by removing some of the Sun's mass.

Projects aiming to domesticate the Sun may well seem quite extravagant today. They are indeed a long way out of reach and far removed from the needs of our present civilisation. Will they ever be seriously attempted? No one can answer this question. I sometimes think of the heros in *Quest for Fire* (*La guerre du feu*) by Jean Rosny Sr, one of my favourite childhood stories. A few thousand centuries ago, our distant ancestors were completely dependent on nature to provide that most valuable of commodities: fire, the ultimate protection against darkness, cold and wild animals. The best they could do was to gather this magical substance wherever and whenever nature offered it: at the foot of a volcano, in a tree struck by lightening, or in a bush fire. It was no easy matter to conserve it and, apparently, they were unable to make fire themselves. In those gloomy times, would they ever have imagined that their descendants might one day master the magical element? And if some of them did dream of that, surely their contemporaries judged this conceptual leap of the imagination to be as crazy as today's projects to tame the solar fire. Of course, we shall never know. But if the human species took several thousand centuries to master fire, since the appearance of the first humanoids, is it unthinkable that after tens of millions more centuries, it should manage to tame the Sun? Personally, I do not think so.

ULTIMATE FUTURES

All that comes into being deserves to perish.

Goethe, *Faust*

And without raising vainly imploring hands to the empty skies, we shall pursue, through indifferent Forces, towards a Future that may match our greatest Dreams, a motion that nothing yet seems compelled to stay.

Jean Perrin

In previous chapters, we explored the possible future of the human species within the local context provided by the Solar System. We concluded that this future was bounded in time, even though the limits could in principle be pushed back a few billion years. We also imagined possible expansions of our civilisation across the Galaxy, without dwelling on the question of its very-long-term future.

In the present chapter, we shall push our investigation to its very limit, embarking upon a voyage to the most distant confines of the future. We shall be led to consider the evolution not only of our own Galaxy, but also of the Universe as a whole. Naturally, we must assume that our means of transport are adequate for such a voyage. In other words, that the basic laws governing the evolution of the Universe are sufficiently well known, and indeed that they do not themselves change in time. These assumptions are far from obvious. The two foundation stones of modern physics, quantum mechanics and general relativity, were unknown only a century ago. Future theories may radically alter our vision of the Universe, as Hubert Reeves has reminded us in the Foreword to this book.

Before beginning our journey into the distant future of the Universe, it is interesting to observe that our vision of the future was long dominated by the idea of Eternal Return. All the great civilisations of the past, from the Babylonians to the Hindus and the Maya, developed their cosmologies on the basis of a cyclic notion of time. After a certain period, which varies in length from one civilisation to another, the Universe regenerates. According to the Babylonians, each cosmic cycle lasted 424 000 years, whilst Hindu mythology referred to much longer periods. For the god Brahma, one day (a *kalpa*) lasts about 4 billion years. When the day draws to a close, all creatures are destroyed and then regenerated. The substance of the Universe, together with Brahma himself, are dissolved into a form of pure Spirit, from which they regenerate every 311 trillion years (3.11×10^{14} years). Greek philosophers in the Stoic school even believed that, during each cycle, the same beings would reappear and the same events occur, an extreme form of Eternal Return that Aristotle could not accept.

This cyclic notion of time was clearly inspired by the observation that many natural phenomena are periodic. Examples like the seasons, phases of the Moon, and so on, were often extremely important for everyday life in the agrarian civilisations of the past. However, with the advent of Christianity, a linear vision of time was introduced, founded on the uniqueness of Christ's death and resurrection. In his book *City of God*, Saint Augustine expressed these ideas with great clarity: 'Christ died once for our sins and, rising again, dies no more.' It is interesting to note that Saint Augustine appeals to the same argument to counter the suggestion that there may be other inhabited worlds, as we saw in Chapter 2.

These two visions of the future, with cyclic and linear times, are also to be found in modern cosmology, as we shall soon see. Although the present state of our knowledge tends to favour a linear time, Eternal Return has not yet been totally excluded. However, our journey into the distant future will no longer be characterised by the question of continuous development and progress, as it was in preceding chapters. These notions will give way to a quest for survival in a Universe that becomes ever more hostile towards the fragile complexity of life.

The long twilight of the gods

Our galaxy is now in the brief springtime of its life – a springtime made glorious by such brilliant blue-white stars as Vega and Sirius, and, on a more humble scale, our own Sun. Not until all these have flamed through their incandescent youth, in a few fleeting billions of years, will the *real* history of the universe begin.

It will be a history illuminated only by the reds and infrareds of dully-glowing stars that would be almost invisible to our eyes; yet the sombre hues of that all-but-eternal universe may be full of colour and beauty to whatever strange beings have adapted to it. They will know that before them lie, not the millions of years in which we measure the eras of geology, nor the billions of years which span the past lives of the stars, but years to be counted literally in trillions.

They will have time enough in those endless aeons, to attempt all things, and to gather all knowledge. They will not be like gods, because no gods imagined by our minds have ever possessed the powers they will command. But for all that, they may envy us, basking in the bright afterglow of Creation; for we knew the Universe when it was young.

Arthur C. Clarke, *Profiles of the Future*

It is difficult to imagine a more poetic description of the long-term future of our Galaxy and other galaxies in the Universe, over time scales hundreds of times longer than anything we have yet encountered in the first three chapters. Indeed our Galaxy, this vast spiral gathering of a hundred billion stars, has already seen its best days. Born something like twelve billion years ago from a condensation of primordial gases, it has witnessed the formation of several generations of stars within its realm. But unlike diamonds, stars are not for ever!

The most massive stars, at least twice as massive as the Sun, shine tens, even thousands of times more brightly than our own star. Unfortunately, their beautiful bold colours, the bright blues and whites, do not last for long. After no more than a billion years, and only a few million years for the most massive amongst them, their hydrogen reserves are already exhausted. Then comes the advanced age of the red giants, of much shorter duration, and finally death, either through the apotheosis of a supernova explosion, or peacefully, by ejecting the stellar envelope in the form of a so-called planetary nebula. In the first case, the fate of stars more massive than ten Suns, the explosion leaves an extremely compact residue of radius less than about ten

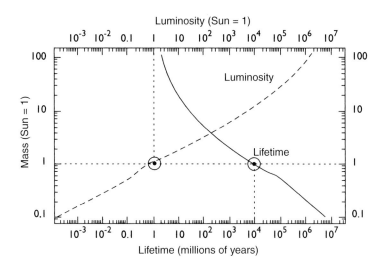

FIGURE 4.1   Lifetimes of the stars (solid curve, lower scale) and luminosities during the core hydrogen burning phase (dashed curve, upper scale), both as a function of the stellar mass. Corresponding values for the Sun are marked with the symbol ⊙.

kilometres. These objects are neutron stars or black holes, which will play an important role in the rest of our story. In the second case, referring to stars of mass less than ten solar masses, the residue is a crystalline object comprising mainly carbon and oxygen. This is a white dwarf, of similar size to Earth. In both cases, however, the main part of the star's initial mass is rejected in the galactic medium. The latter thus grows gradually richer in heavy elements (carbon, oxygen, iron, etc.) produced by nuclear reactions that took place inside the stars.

These bright stars, so active in nucleosynthesis, are not very abundant, representing only a few per cent of any stellar generation. Stars of average mass, between half and twice the Sun's mass, are far more common. These stars can shine for billions, or even tens of billions of years, with orange to bright yellow colours. Like the Sun, they end their lives as white dwarves, after passing through the red giant stage. Despite low individual luminosities, their collective brightness dominates the Galaxy at present due to their superiority in numbers. Indeed, unlike their relatives in the first category, few of these have died since the birth of the Milky Way.

There remains a third class containing the smallest stars, those with masses less than half the Sun's mass. They are tens or thousands of times less luminous than the Sun and their pale red colours can barely be made out on the sky background. However, parsimonious management of their low energy reserves allows them to survive for hundreds of billions of years. The smallest of them, with just a tenth of the Sun's mass, will live more than ten thousand billion years. This is clearly the case for our nearest neighbour, Proxima Centauri, which is destined to live a thousand times longer than our own star. Unlike the Sun and other typical stars, these red dwarves will never become giants and their luminosity will never increase by more than a factor of ten relative to present values. It is this species, more numerous than the two others put together, that will inherit the Galaxy in the very long term.

Today, the Milky Way has almost exhausted its resources. After 12 billion years of evolution, it has transformed 90% of its gas into stars, many of which are already dead. As the large and bright blue stars disappear, the luminosity of the Galaxy diminishes and its colours become less bright, as it gradually becomes dominated by solar-type stars. Of course, further generations of massive stars will continue to enter the galactic scene, but never enough to balance the losses, since gas reserves have been seriously depleted. Slowly, inexorably, our aging Galaxy turns to an orange–yellow hue. In a few tens of billions of years from now, almost no more gas will remain and its ability to form stars will come to an end. The Milky Way will then look like its tired relatives, the elliptical galaxies. Their reddish hue and absence of gases or massive stars is witness to a wild youth, too lively to sustain.

Before reaching that epoch, the relatively peaceful existence of our own Galaxy will certainly be affected by its near neighbours. The largest galaxy in our cosmic neighbourhood is Andromeda, about 2 million light-years away. Andromeda is presently approaching the Milky Way at a speed of 120 km/s. Calculations show that the trajectories of the two galaxies will almost meet in 6 billion years from now, just as our Sun begins to swell up in its red giant phase. During the close encounter, Andromeda's gravitational field will act on gases in our own Galaxy (especially those situated on the periphery of the disk), inducing a wave of star formation. This phenomenon can already be observed in the case of other interacting galaxies. Pursuing its wild course across intergalactic space, Andromeda will move away for a certain time. However, it will return to make closer and closer approaches, for the two galaxies are bound by their mutual gravitational attraction. After a few hundred billion years, Andromeda, the Milky Way and the dozen or so small galaxies making up our Local Group will finally merge together into a single vast stellar system. Contrary to what we might imagine, this coalescence of galaxies will not be a spectacular affair, even if it occurs at exceptionally high speeds. There is so much space between the stars in a galaxy that the probability of a collision between two stars is quite negligible.

The long galactic twilight will then begin. Civilisations existing in this epoch will have to work miracles to prolong their existence every time their host star approaches its end. Moving to another, still active star or artificially extending the lifetime of their own dying host by reducing its mass are survival techniques we have already discussed in the last chapter. The astro-engineers of this distant future, quasi-gods in relation to ourselves according to Clarke, will doubtless find other methods. For example, they might use gases extracted from dying stars to fuel controlled thermonuclear fusion reactors, or even to create new stars. A way must then be found to stock this gas over long periods in a rarefied form, very likely using magnetic fields. Then at the right moment, they could switch off the magnetic field and let the gas cloud condense under its own weight. When the temperature at the centre of the cloud reaches a few million degrees, hydrogen fusion reactions will begin, and a new sun will make its appearance. Moreover they might use the vast reserves of hydrogen remaining intact in brown dwarves to fuel their thermonuclear fusion engines. A brown dwarf is a body at least ten times lighter than the Sun which is unable to burn its own hydrogen reserves, because of its insufficient internal temperature. In order to face a shortage of heavy elements (carbon, oxygen, nickel, iron, and so on), the future astro-engineers could first merge together several brown dwarves to make a massive star, and then induce its explosion in a supernova, recovering the ejected materials. The explosion of supernova SN1987A in the nearest galaxy to us, the Large Magellanic Cloud, on 23 February 1987, threw out a mass of nickel equal to twenty thousand times the mass of our own planet.

Such astro-engineering exploits will seem rather commonplace to any civilisation capable of administering the energy and material resources of a whole galaxy (type III in Kardashev's classification). But despite their unimaginable powers, future civilisations will not be able to escape the inevitable indefinitely. After a few thousand billion years, they will find themselves with no further source of stellar energy, in a Galaxy made up entirely of cold objects: brown dwarves, black dwarves (completely cooled former white dwarves), neutron stars, black holes, planets and asteroids. The wretched darkness of the long galactic night will be broken only by the occasional infrared glow of one of these cool objects.

The heat death of the Universe

Nineteenth-century science was not unaware of the long twilight awaiting the Universe and the prospect that energy might be lacking on the cosmic scale in some very distant future. It was thermodynamics which engendered this awareness. The initial objective of thermodynamics was to understand heat and mechanical energy exchanges in steam engines. As it developed, rather general concepts emerged with applications to almost all physical systems. The strength of thermodynamics lies precisely there, in the generality of its concepts, which are independent of whatever detailed structure a system may have. These concepts are thus endowed with a certain universality.

The universal scope of thermodynamics is symbolised by its famous second law, linked historically with the names of the French engineer Sadi Carnot and the German physicist Rudolf Clausius. Carnot studied the maximum efficiency of machines transforming heat into work, recognising that the crucial factor is the temperature difference between the heat source and the indispensable 'cold reservoir'. Clausius observed that in any operation of this kind, there is an unavoidable loss of energy in the form of heat, which dissipates into the environment and subsequently becomes unusable. The original formulation of the second law of thermodynamics by Lord Kelvin was based precisely on this remark: 'It is impossible to convert a given quantity of energy into work with one hundred per cent efficiency.'

However, the second law is better known according to an alternative formulation by Clausius, who also introduced the notion of entropy (from the Greek *entropia*, meaning an intrinsic tendency to change). In the case of heat engines, we may say that entropy measures an inability to obtain work. Naturally, it depends on the temperature difference between the heat source and the cold reservoir. When this difference is great, the system has low entropy (hence, a large capacity for doing work), conversely with a small difference a system has high entropy. For example, the high temperature of the Sun makes it a source of low entropy and hence eminently useful to ourselves. In contrast, even though the oceans contain a vast quantity of energy, they are useless from an energetic point of view, lacking the indispensable cold reservoir. Since heat spontaneously transfers from hotter to colder bodies, temperature differences always tend to level out. Whence Clausius' formulation of the second law: 'The entropy of an isolated system can only increase as time goes by.'

According to another interpretation, the entropy of a system is a measure of its disorder. An organised and well-structured system has low entropy. But left to itself (i.e., without interacting with its environment), it tends to become disorganised, its structures dissipating and drifting towards uniformity. When a drop of milk disperses in our coffee, or an old house falls to ruin, we are witnessing examples of this spontaneous increase in disorder. Of course, new structures may appear: the old house may be renovated or a new one built. The entropy is then reduced locally, but at the expense of a general increase within some larger system.

Can we conceive of a violation of the second law? In 1867, Scottish physicist James C. Maxwell flirted with the idea. He considered a box filled with a gas at uniform temperature and pressure (in thermodynamic equilibrium, to use physicist's jargon). Heat flow being impossible within the system, this state corresponds to maximal entropy. Now imagine a tiny being, referred to as Maxwell's demon by Lord Kelvin, who is able to intervene in the following way. By opening and closing a diaphragm in the middle of the box, it allows high-speed gas molecules to pass through from one side, but not from the other; likewise, it allows low-speed molecules to go the other way (and only the other way). After a certain time, all the high-speed molecules are located on one side of the diaphragm, and all the low-speed molecules on the other. As the temperature of a gas depends on the average speed of its constituent molecules, we would have a temperature difference between the two compartments, and useful energy could subsequently be extracted! Of course, the entropy of the gas would then be less than its initial value. This was no doubt a paradox, but Maxwell preferred to give a statistical interpretation at the time.

A more detailed examination of the paradox by Hungarian physicist Leo Szilard in the 1920s showed that the demon, no matter how intelligent it might be, would never achieve its aims. In order to know the speed of the molecules, it would have to illuminate them, rather as police radars are used to detect speeding motorists. This act of measurement involves an energy expenditure, and hence an increased entropy inside the box, easily sufficient to balance the reduction obtained by the demon's sorting process. Incidentally, this solution to the paradox illustrates a third, more modern interpretation of entropy as a measure of the lack of information we have on a system.

The failure of Maxwell's demon instantiates the inescapable second law of thermodynamics. Its scope is so great that Sir Arthur Eddington, the greatest astronomer between the wars, described it as the supreme law of nature: 'The law that entropy always increases – the Second Law of thermodynamics – holds, I think, the supreme position among the laws of Nature. If someone points out to you that your pet theory of the Universe is in disagreement with Maxwell's equations – then so much the worse for Maxwell's equations. If it is found to be contradicted by observation – well, these experimentalists do bungle things sometimes. But if your theory is found to be against the Second Law of thermodynamics I can give you no hope; there is nothing for it but to collapse in deepest humiliation.'

Invoking this universal applicability of the second law in 1854, German physicist von Helmholtz envisaged the future Universe in a state of absolute uniformity, in which all temperature differences would reduce to zero and a global thermodynamic equilibrium would be reached. The idea of the heat death of the Universe had been born. Ten years later, Clausius stated Helmholtz's ideas explicitly: 'The more the Universe approaches the limiting condition in which the entropy is a maximum, the more do the occasions of further changes diminish; and supposing this condition to be at last completely obtained, no further change could evermore take place and the Universe would be in a state of unchanging death.'

These ideas were the subject of considerable debate during the second half of the nineteenth century and had a great impact on philosophical visions of the world at the time. Moreover, they were not unknown to the romantic tradition which dominated over the whole of this period. Even so, renowned scientists such as the Frenchman Henri Poincaré and the Austrian Ludwig Boltzmann attempted to contest the absolute validity of the second law, attributing a merely statistical import to it. Boltzmann thus suggested that the increased entropy in our region of the Universe must be compensated for by a reduction elsewhere, in such a way that the mean entropy should remain constant. This differential behaviour would result, in Boltzmann's view, from a difference in initial conditions. In the past, he claimed, our part of the Universe found itself in a low entropy state, whilst other parts began with a high entropy, which was supposed to diminish in some natural way. Poincaré put forward the same statistical reasoning, although in a slightly different context. In his view, since the Universe is made up of a finite number of constituents, they can only form a finite number of different configurations. These constituents must necessarily return infinitely often to a given state, even if an incredibly long time is needed for each recurrence. In this sense, our presence within the current phase of decay in the Universe is a mere question of chance.

The notion of the heat death of the Universe naturally incited a reaction from Friedrich Engels, co-founder of the socialist movement with his friend Karl Marx in the middle of the nineteenth century. The pessimistic implications of the idea were in flagrant disagreement with the notions of evolution and progress, key features in the dialectic materialism that the two philosophers professed. In his book *Dialectics of Nature*, Engels attempted to exorcise the spectre of heat death, using arguments similar to Poincaré's:

> ... an eternally repeated succession of worlds in infinite time is the only logical complement to the co-existence of innumerable worlds in infinite space ... It is an eternal cycle in which matter moves ... We have the certainty that matter remains eternally the same in all its transformations, that none of its attributes can even be lost ...

The German philosopher Friedrich Nietzsche also used arguments of this kind to revive the myth of Eternal Return at the end of the nineteenth century. In fact, Nietzsche devoted several years to studying the physics of his day so that he could defend his thesis. The conclusions he drew form the basis for his philosophical work. According to Nietzsche, Eternal Return implies that the idea of progress is null and void, and that life has no meaning (nihilism); or again that God does not exist, and that if he does exist, he is as absurd as the Universe he has created.

Nietzsche's ideas had a profound effect on twentieth-century philosophy. Among others, they inspired historians Oswald Spengler and Arnold Toynbee, who proposed a cyclic vision of human history. Rather than a continued progress, history was a repetition of the basic cycle: birth, rise, decline and death of every civilisation. In the last chapter, we saw how Olaf Stapledon adopted this vision of history in his novel *First and Last Men*. Eternal Return is also the basis for Albert Camus' *The Myth of Sisyphus*, a major contribution to existentialist philosophy. In Greek mythology, the gods condemned Sisyphus to push a rock up to the top of a mountain in hell. This is a truly terrible punishment, because he knows that all his efforts are in vain. When he reaches the summit, the rock will fall back down and he must start all over again. Camus nevertheless finds that humanity can overcome the absurd conditions of its existence and hence become the master of its own destiny.

Despite this comeback of the Eternal Return, the idea of the heat death of the Universe was predominant at the end of the nineteenth century, at least in scientific circles. It was only in 1914 that the first valid arguments came out against this notion, stated by the French physicist and philosopher Pierre Duhem: '[The heat death hypothesis] implicitly assumes the assimilation of the Universe to a finite collection of bodies isolated in a space absolutely void of matter; and this assimilation exposes one to many doubts ... It is true that the entropy of the Universe has to increase endlessly, but it does not impose any lower or upper limit to this entropy; nothing then would stop this magnitude from varying from minus infinity to plus infinity while the time itself varied from minus infinity to plus infinity.' We shall see shortly that Duhem's criticism is highly relevant in the context of relativistic cosmology, although the latter cannot guarantee an infinite prolongation of life.

Decay or evolution?

By the end of the nineteenth century, it had become clear that universal decay as predicted by the second law of thermodynamics was in contradiction with biological evidence. Darwin's theory of evolution had revealed a steady complexification of living matter. The difficulty in reconciling the two visions of the world led some quite simply to deny that the second law was applicable to living systems. Amongst these was the French philosopher Henri Bergson: 'Life ascends the slope that matter descends.' In his book *The Ghost in the Machine*, Arthur Koestler was even more radical:

> Clausius' famous second law of thermodynamics asserted that the universe is running down, like a clock affected by [metal] fatigue, toward the Cosmic Heat Death ... Only in recent times did Science begin to recover from the hypnotic effect of this nightmare, and to realize that the second law applies only in the special case of so-called closed systems ... that this law did not apply to living matter and was in a sense reversed in living matter, was indeed hard to accept by an orthodoxy still convinced that all phenomena of life could ultimately be reduced to the laws of physics.

The Jesuit palaeontologist Pierre Teilhard de Chardin also adopted a resolutely optimistic stance. His view of the relationship between science and religion was not approved by the hierarchy of the Catholic Church and his philosophical works were not published until after his death in 1955. For Teilhard de Chardin, there were two worlds, one physical and the other spiritual, each having its own form of energy. In the physical world, energy downgrades according to the laws of thermodynamics, whereas energy in the spiritual world is not subject to these laws, but progressively complexifies matter according to the divine scheme. The development of intelligence, or Noogenesis, will eventually culminate at the so-called Omega point where cosmic intelligence becomes identified with the spirit of God.

> For a whole century, physical science was dominated by the idea that energy dissipates and matter decays. Challenged by biology, it began to understand that, in parallel with this decay, a second process is developing in the Universe, as general and fundamental as the first. I am referring to the progressive concentration of physico-chemical elements into more and more complex forms, each step being accompanied by a more advanced form of spiritual energy. The ebbing flow of Entropy is thereby equalled and even exceeded by the rising tide of Noogenesis.

In opposition to Koestler and Teilhard de Chardin's ideas, physics allows no exception to the second law. Living systems constitute only an appearance of exception. They are not closed systems and their behaviour, considered as part of some larger system, complies perfectly with the demands of thermodynamics. For example, the development of life on Earth represents an astonishing establishment of order, and hence a local decrease in entropy. However, it is only possible through the help of energy from the Sun, whereby it contributes positively to the entropy of the Earth–Sun system. The high tide of Noogenesis cannot dispel the spectre of heat death.

The moving vision of an extremely slow death in a remote future we cannot determine never really inspired science fiction writers, who were more attracted by spectacular cosmic disasters. Among the rare exceptions, Isaac Asimov deals with the problem in a quite magnificent way in his short story *The Last Question*, written in 1955. The tale begins with the inauguration of a world supercomputer, built at the dawn of the twenty-first century to administer the entire worldly needs of Earth's inhabitants. One of the engineers, under the influence of alcohol, defies the machine with the following impossible question: 'Is it possible to reverse the entropy of the Universe?' The computer replies: 'I have not sufficient information to answer.'

The centuries and millennia pass, and both humans and supercomputer evolve. From time to time, someone asks the computer the same question, invariably obtaining the same response: 'I have not yet sufficient information available.' As time goes by, the question acquires practical significance, and even becomes urgent, for the last stars are dying and intelligence is entering a long death struggle. Eventually, its last representative passes away, having posed the famous question one last time, and obtaining the same reply from the cosmic supercomputer, set up somewhere in superspace. But the supercomputer is still running (by which miracle, Asimov does not specify) and finds itself constrained to execute its masters' last order. Over the centuries, it continues to gather the required data, patiently studying and analysing it, until one day it is finally ready to pronounce the famous phrase: 'Let there be light! And there was light.' This is indeed a sublime ending to the best story of its kind, the good doctor Asimov's own favourite. It is impossible, at the present stage of our knowledge, to escape the hold of the second law other than by a twist of the type used here.

Nineteenth century physics could not determine the date at which the heat death of the Universe would occur, because it was not known how the stars obtain their energy, nor how the galaxies evolve. With this knowledge they might have situated the event within the next few thousand billion years. But the prediction would have been wrong. In contrast to the nineteenth century conception of the world, the Universe is not static. Advances made in the twentieth century allow us to imagine radically different perspectives for the future evolution of the Cosmos, although they remain just as pessimistic as the ones they supplant with regard to the fate of life.

The Universe of the Big Bang

One of the greatest philosophical impacts of twentieth-century science concerns our vision of the Universe. A series of theoretical and observational discoveries, begun in the 1920s, literally shattered the ancient idea of a static, immutable and eternal Cosmos, replacing it by a Universe in permanent evolution.

The history of modern cosmology, a discipline devoted to the study of the Universe as a whole, began with the development of general relativity by Albert Einstein in 1915. This is indeed the best theory available today for the description of gravity, the only relevant force on cosmic scales. Whilst studying the equations of relativistic cosmology in 1922, the Russian mathematician Alexander Friedmann noticed something quite extraordinary. The Universe cannot be static, but rather it must partake of an overall motion, either contracting or expanding. Einstein's philosophical objections were swept away in 1929, when the American astronomer Edwin Hubble discovered that all galaxies are moving away from our own with speeds proportional to their distance from us. The model of the expanding Universe was born. It should be noted that we can in no way deduce a central position for ourselves from these observations, just as an ant on an expanding balloon cannot claim to be in a privileged position on its surface.

The image of expansion naturally suggests that, in the past, the Universe was much denser than it is today, and consequently much hotter, according to a well known property of gases. (Note that the 'atoms' of the cosmic gas are the galaxies themselves.) Extrapolating this idea to its very limit in the 1930s, the Belgian priest Georges

Lemaître suggested that the Universe had sprung from an extremely hot and dense initial state, the 'primeval atom', several billion years ago. At the end of the 1940s, the American physicist (of Russian origin) George Gamow and his collaborators investigated Lemaître's 'primeval atom' using nuclear physics. Their work led to two important conclusions.

Firstly, if the primordial Universe had ever had a temperature greater than a few thousand degrees, a residual radiation must remain today from its flamboyant youth. This relic, invisible to our eyes, would correspond to a temperature of just a few degrees above absolute zero, because of cooling caused by the intervening expansion.

Secondly, if the Universe had ever witnessed even higher temperatures, of order a few billion degrees, matter could only have existed in the form of elementary particles, such as protons, neutrons, electrons, and so on. Nuclear reactions between such particles in the heat of the primordial Universe would have produced some light nuclei, in particular, helium-4, which is the most abundant chemical element after hydrogen on the cosmic scale.

Those predictions were subsequently almost forgotten by physicists for around fifteen years. But in 1965, the American engineers Arno Penzias and Robert Wilson noticed that the new radio antenna they were testing at the Bell Laboratories in New York was detecting a mysterious background radiation. It appeared to be coming from all directions in the sky with equal strengths. The theoreticians reacted immediately. This was exactly the thermal background radiation Gamow had predicted, a relic from the early youth of the Universe, which has now cooled down to just 3 K ($-270\,^{\circ}$C) after 15 billion years of expansion. This discovery convinced the scientific community that the Lemaître–Gamow model was right. Further support was provided in 1964 by the British astrophysicist Fred Hoyle and his colleague Roger Tayler when they showed that the stars, the source of almost all chemical elements, could never have produced the high cosmic abundance observed for helium-4 (about 25% by mass). Only one other possible source remained: the hot primordial Universe.

Since the middle of the 1960s, the Big Bang theory, as it was ironically christened by Fred Hoyle in a BBC radio programme in 1948, has prevailed as the only model capable of explaining three fundamental observations: the retreating galaxies, the cosmological background radiation and the abundance of helium-4 and other light elements such as deuterium. Within the context of this model, astrophysicists have been able to retrace the history of the Universe, rather as palaeontologists elaborate their evolutionary theories on the basis of fossils they have discovered. In this universal history, there are two great periods.

The *radiation era* lasted around 300 000 years after the Big Bang. During this period, the Universe remained undifferentiated, a kind of soup of radiation, elementary particles and light nuclei which gradually became dilute and cooled. The entropy of this uniform mixture would always have been close to the maximal value, no structure formation being possible within it. Little by little, the soup was transformed into a thick white fog, which became yellow and then red, whilst still remaining opaque to electromagnetic radiation thanks to the many collisions between photons and charged electrons. When the temperature fell to a few thousand degrees, the first atoms formed as nuclei began to capture electrons. For the first time, the medium became transparent to photons, since they interact only weakly with electrically neutral particles such as atoms. Since then, these photons, bearing witness to the hot primordial Universe, have propagated imperturbably in straight lines, apart from the occasional collision with some heavenly body (or with Penzias' and Wilson's antenna).

The *matter era* concerns the whole of the remaining history of the Universe. The cosmic gases began to condense here and there as a result of local gravitational effects, thereby removing themselves from the global expansion of the Universe. Amplified by their own weight, these condensations were the seeds of the first galaxies. Within them, even smaller condensations formed that were to be the first generations of stars. Of course, entropy was reduced locally as each of these structures (galaxies, stars, planets, and eventually, life) formed. However, this reduction was accompanied by emission of electromagnetic radiation (heat), in such a way that the entropy of the whole Universe has always been on the increase.

This is the scenario cosmologists have produced over the past quarter of a century to tell the story of our Universe. It has stood up well to observational evidence, a success which is in no way affected by the fact that certain aspects, such as the detailed mechanism of galaxy formation, are still relatively obscure. As far as the origin of the Universe is concerned, it still evades our comprehension. Indeed, we may not even approach it as closely as we would wish. Current physics allows us to go back as far as the time when the temperature was 10^{32} K (100000 billion billion billion K). Above this temperature, reached around 10^{-43} seconds after the Big Bang (using a false chronology, in which zero time corresponds to infinite temperature), general relativity no longer applies. A new theoretical tool is needed. However, this tool, which goes by the name of quantum gravity, is not yet available. And until it has been established, the temperature 10^{32} K will remain the limit of our understanding of the primordial Universe.

Physicists already have some ideas for going beyond the Planck limit (named after the founder of quantum physics). Amongst these ideas, based on the most recent speculations in the physics of the infinitely small, chaotic inflation is one of the most attractive from a philosophical point of view. Developed by the Russian physicist Andrei Linde in the 1980s, this theory views the observable Universe as a tiny portion of a gigantic bubble which itself measures 10^{3000} light-years across. The bubble inflated to these colossal dimensions in just a tiny fraction of a second after the Planck time, according to the inflationary universe scenario developed by the American cosmologist Paul Guth in 1980. In Linde's theory, the bubble emerged from a quantum fluctuation of the vacuum, that ethereal substrate that fills the whole of space and that would appear to have quite extraordinary properties under the extreme conditions reigning at the Planck time. Other bubble universes would also have emerged in a chaotic way from fluctuations in this substrate, each one with different physical properties. Some would already have collapsed, restoring their matter and energy to the vacuum from which it was borrowed. Others would still be pursuing their expansion at their own rate, separately creating their own space–time and never interfering with their neighbours. The impossibility of any physical contact between the bubble universes greatly reduces the physical interest of Linde's theory, despite its undeniable philosophical attraction.

Before tackling the predictions of modern cosmology with regard to the future evolution of the Universe, we must tarry a moment on a fundamental paradox, which will not have escaped the reader's attention.

Entropy and gravity

Within the undifferentiated and homogeneous soup of the primordial Universe, entropy was apparently very close to its maximal value. The

Universe was stillborn from the thermodynamic point of view, since only temperature gradients can generate useful work. How then can we explain the appearance of structure within the cosmic plasma? What miracle allowed temperature differences to build up between the stars and the rest of the Universe? And how did the entropy manage to increase still further when it was already close to its maximal value right at the beginning?

The answers to these questions are far from trivial. However, they are extremely important if we are to understand the future prospects for our Universe. The notions involved are rather abstract, making this section somewhat more difficult than the rest of the book. But we can already give the answer in one word: gravity.

The familiar example of life on Earth will provide us with an insight into the problem. We know that life is only possible because of energy radiated by the Sun and intercepted by Earth's surface. Clearly this energy is not stored by Earth, nor by living organisms, otherwise they would heat up considerably! It is simply used to maintain the complexity of life at a high level (or to hold down its entropy). Once used, the energy is radiated away in the form of heat, and it is here that the inevitable increase in the system's entropy must occur. Indeed, the solar energy arriving on Earth has low entropy. It takes the form of visible photons whose average energy corresponds to a temperature of 5800 K, which characterises the Sun's surface. When it leaves Earth, at a temperature of only 300 K, the same energy takes the form of infrared photons, which are twenty times less energetic. Hence, twenty photons are reradiated by Earth for each photon received from the Sun. This multiplication in the number of photons increases the disorder and hence also the entropy of the Earth–Sun–space system. As a heat engine, Earth transforms low-entropy energy into high-entropy energy which is thereafter much more difficult to use.

Why is the Sun a source of low entropy? Quite simply because it is so much hotter than its environment. And it is hotter because its high gravity forced the dispersed gases of the protosolar cloud to contract and increase its temperature, in accordance with a well-known property of gases. We thus arrive at a basic observation which goes against nineteenth-century physical knowledge and our own intuition. If we ignore gravity, the maximal entropy state of a gas is the dispersed isothermal state in which the gas tends to have the same temperature everywhere and to occupy the whole available volume. When gravity is taken into account, the situation is different. The dispersed gas has low entropy. It can increase it by condensing locally, even though this creates internal temperature differences.

Therefore, if we consider both matter and its gravitational field, we can understand why the Universe was able to structure itself. From a gravitational point of view, the initial dispersed gas state had very low entropy, which could further increase. Of course, during the whole of the radiation era, the gas was still too hot to condense. Thermal agitation of its constituents vigorously opposed the hold of gravity. But once expansion had calmed this agitation down by the cooling effect it engendered, gravity was finally able to overcome. Structures were created which defied the second law of thermodynamics on a local level: stars, planets and life itself. This extremely simplified picture of the emergence of complexity resolves the paradox mentioned at the beginning of the present section, viz., the question of how the Universe could have evolved when it was already thermodynamically dead at its very birth.

Furthermore, it turns out that the expansion of the Universe causes entropy to increase forever, without ever reaching a maximal value. Indeed, the maximal value of the entropy in an expanding Universe increases with time, because the greater the volume of a system, the higher its maximum possible disorder. (To illustrate this volume effect, we can imagine our bookshelves in a state of total disorder, and then imagine the same number of books scattered around Wembley Stadium. If we measure the disorder by the time required to tidy up all the books, then it is much greater in the second case.) In order to understand why the entropy of the expanding Universe cannot attain its maximal value, the example of a swimming pool with moving walls is sometimes cited. If we attempt to fill the pool with water, the volume of water can increase continuously without ever running the risk of overflowing. The static Universe of the nineteenth century would be a swimming pool with fixed walls. Sooner or later it must overflow. Hence, modern relativistic cosmology has borne out Pierre Duhem's reasoning when he opposed the heat death of the Universe, as we saw in a previous section.

These notions were mainly developed over the past half century. They are foreign to the mechanistic thinking that prevailed in the nineteenth century, and may encourage us to indulge in a certain optimism as regards the distant future of life in the Universe, for they seem to relegate heat death to the past. This is the opinion of Hubert Reeves, among others, expressed with great force in his book *Hour of our Delight: Cosmic Evolution, Order and Complexity*. The roles played by gravitation and entropy in the emergence of cosmic complexity are very clearly presented here.

However, it transpires that heat death may be present both in the past and in the future, in ways that would have been quite beyond the imagination of nineteenth-century physicists. It is highly probable that the wonderful complexity which surrounds us is merely a brief reprieve: sublime but destined to disappear in the long term.

The dark side of matter

The origin of the Universe lies hidden from relativistic cosmology at the present time. Its future, on the other hand, proves to be more accessible. The fate of the Universe depends on a struggle between the impetus of the primordial explosion and the omnipresent effects of gravity which tend to decelerate the expansion. As gravitational strength is directly proportional to the mass involved, we see that the destiny of the Universe depends on its mass, or rather, its average density (the mass per unit volume). Solution of the cosmological equations reveals two possible cases:

- If the density of the Universe is less than (or equal to) a certain critical density, expansion will continue *ad infinitum*, albeit decelerated, gravity being too weak to bring it to a halt. The Universe is then said to be open and infinite in space and time.
- If the density of the Universe is greater than this critical value, expansion will slow down and finally stop one day, before reversing into a contraction that will become ever faster as it lurches back to a state similar to the Big Bang. The Universe is then closed, and spatially and temporally finite.

Solution of the cosmological equations also provides us with the critical value of the density mentioned here. It is time dependent. Today, around 15 billion years after the beginning of the expansion, it

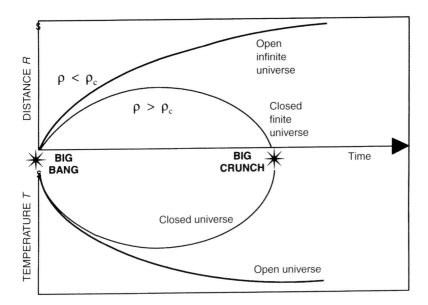

FIGURE 4.2   According to modern cosmology, the future of the Universe
depends on its density. If it has density greater than a certain critical
value ρ_c, expansion will stop and the Universe will proceed to contract
down to a very hot, dense state known as the Big Crunch. In the opposite case, expansion and its accompanying cooling will continue forever.

is about 10^{-29} grams per cubic centimetre, or the equivalent of
3 protons in a cube of side one metre. But what in fact *is* the present
density of the Universe? The density of matter which can be observed
directly in a luminous form, mainly inside galaxies, is evaluated at
slightly less than one hundredth of the critical value. It would thus
seem that the Universe is open by a clear factor of one hundred.
However, the situation is not so simple. Indeed, a series of key observations would indicate that most of the mass of the Universe has still
escaped detection. This is not a recent idea, but it has become one of
the major topics of research in astrophysics over the past two decades.

The Coma cluster, situated about 300 million light-years from us,
contains several thousand galaxies. In 1933, the Swiss astronomer Fritz
Zwicky noticed that the galaxies in this cluster are moving at high
speeds, of the order of several hundred kilometres per second. At such
speeds, the galaxies should have escaped from the gravitational attrac-

tion of the cluster long ago. Zwicky concluded that the cohesion of the Coma cluster must be explained by the presence of huge amounts of invisible matter. This was the first indication of abundant dark matter on an extragalactic scale, but it lay neglected by astronomers for many years. Later observations have confirmed this conclusion for most clusters of galaxies.

On an even greater scale we find superclusters of galaxies, the largest concentrations of matter in the Universe, made up of several clusters and groups of galaxies. Our local group contains about twenty members and is located at the edge of our local supercluster. At the centre of this supercluster lies the Virgo cluster, some 50 million light-years away. In the 1980s, astronomers observed that the motion of the local supercluster, and also of the neighbouring Hydra and Centaurus superclusters, does not quite conform to the global expansion of the Universe. Rather it seems to be perturbed by something very massive. This mass concentration, evaluated at about thirty times the mass of a supercluster, would appear to be hidden behind the Virgo cluster, at a distance of 150 million light-years. Astronomers have christened it the Great Attractor.

Dark matter has also manifested its presence on the scale of individual galaxies. Since the 1970s, astronomers have realised that spiral galaxies rotate more quickly than their luminous mass would allow. There must therefore be much more invisible matter to ensure their cohesion, otherwise the outer regions of these galaxies would long have been ejected into space by centrifugal forces. This observation applies equally to our own Milky Way.

There are therefore clear indications for the existence of dark matter on all scales, from individual galaxies, through groups and clusters of galaxies to superclusters. Astronomers have even noticed that the ratio between dark matter and luminous matter increases on larger scales. On the scale of spiral galaxies, the ratio is of the order of 10, but can rise to around 30 in groups and clusters of galaxies, and as much as 100 on the largest scales yet explored. As the density of luminous matter is about 1% of the critical density, we see that dark matter could be sufficient to close the Universe if its density is as high as these large-scale observations would suggest. Unfortunately, the uncertainty in these measurements is so great that no conclusions can be drawn as yet. In fact, all the observations presently available suggest that the density of the Universe (including both visible and invisible matter) would in no case exceed half the critical density.

Apart from observations, there are two theoretical arguments that can help us to probe the density of matter in the Universe, the first fairly reliable and the second much less so. According to the scenario of primordial nucleosynthesis, the abundance of the light nuclei produced in the hot, early Universe depends on the density of ordinary matter, for it is this which determines nuclear reaction rates. By measuring the abundance of light elements such as helium, deuterium and lithium in the oldest stars and remote galaxies, this density can in principle be determined. Using this method, it is found that the density of ordinary matter can in no case exceed 10% of the critical value. This rather reliable argument shows that our familiar matter (atomic nuclei making up stars, planets and galaxies) is not enough to reverse the expansion of the Universe.

The second argument is based on the inflationary model of the Universe, mentioned briefly in a previous section. This theory makes use of discoveries in modern microphysics to explain some observable properties of the Universe that have no satisfactory interpretation in the context of the standard Big Bang model. Despite certain attractive features, there is as yet no clear observational support for the inflationary model of the Universe. The only firm prediction that might one day be confirmed by observation is that the cosmic density must exactly equal the critical value. In this case the Universe, marginally closed or open, will continue its expansion for all eternity. Referring to arguments based on primordial nucleosynthesis, which do not allow the density of ordinary matter to exceed 10% of the critical density, we are then forced to conclude that 90% of the mass of the Universe is in some exotic form. The theories of modern microphysics can propose a wealth of candidates, particles with strange names like axions, photinos, and many more. Produced in large numbers right at the beginning of the Universe, these particles would not have affected primordial nucleosynthesis and yet today would dominate the dynamics of the Universe by their collective mass. At present no observational evidence has been mustered to confirm their existence. However, absence of evidence is not evidence of absence. For the sake of completeness, we must consider both possible cases, an open or a closed Universe, when imagining the distant future of the Cosmos.

Up to the final collapse

The future of a closed Universe is well determined, although its duration is uncertain because it depends on the density. If the density exceeds the critical value by 10%, expansion will continue for a further 350 billion years or so, corresponding to about twenty times the present age of the Universe. By this time, only stars with masses less than one quarter of the solar mass will still be alive. Their faint red glow will barely light up the dying galaxies. The temperature of the cosmic background radiation will have fallen to 0.3 K, whilst the galaxies, carried by universal expansion, will be situated ten times further away from one another than they are at present. All these numbers will be considerably greater if the density exceeds the critical value by only 1%. In this case, expansion will cease after about 20 000 billion years in a completely dark Universe. No star will have survived this long (except possibly artificial stars created to satisfy the energy requirements of type II or type III civilisations).

There will be nothing whatever to indicate to survivors (always supposing there are any!) that maximum expansion has been reached. But gradually, by observing distant galaxies, they will realise that a major change has taken place. Indeed, during the whole of the expansion period, radiation from the receding galaxies reaches us with more stretched wavelengths than if it had originated in galaxies at rest. This shift towards the red (since red light has the longest wavelength in the visible part of the spectrum) was just the feature which in 1929 allowed Edwin Hubble to conclude that the galaxies were receding. When the contraction stage begins, the galaxies will begin to move back towards one another and their light will reach us with shorter wavelengths. At first, this blueshift (blue corresponding to the shortest wavelengths in the visible spectrum) will be observed only in the nearest galaxies, and then in more and more distant galaxies as time goes on.

The contraction phase of the Universe will be symmetric with the expansion phase and of almost the same duration. At first, the Universe will contract slowly, without causing any major events. The temperature of the cosmic background radiation and the energy of its photons will very slowly increase. Little by little, the extremely rarefied 'gas' of galaxies will also begin to feel the effects of the contraction. Its basic components will acquire a more and more agitated motion, like the atoms in a compressed gas.

Clusters currently occupy about 1% of the volume of the Universe. When this volume has been reduced by a factor of one hundred, they will merge with their neighbours and cease to exist as distinct structures. At this epoch, a billion years before contraction comes to an end, the temperature of the cosmic radiation will still be relatively low, somewhere around 30 K. The galaxies, independent from this point on, will be moving around at speeds of 500 km/s, whilst still taking part in the overall contraction of the Universe.

Collapse will continue and 900 million years later, the size of the Universe will have been reduced by a further factor of one hundred, so that its volume will be only one millionth of what it is today. Galaxies will then suffer the same fate as the clusters. They will merge together and their constituent stars, endowed with speeds of around 3000 km/s, will now become the atoms of the cosmic gas. The cosmic radiation temperature will have reached 300 K, greater than the average temperature across the surface of our planet at the present time. Space will no longer be the cold place it is today and our distant descendants will be able to move around in it without wearing any thermal protection to keep warm. However, they will encounter more and more difficulty evacuating the entropy produced by their own metabolism. Space will become an ever greater source of heat, stifling all life forms in the long run (at least, those we know about today). It should not be imagined that space will be any less dark than it is today (for eyes analogous to our own, at any rate). The characteristic frequency of photons making up the cosmic background radiation will still lie in the infrared region of the spectrum.

The situation will continue to evolve. Seventy million years later, the Universe will be a thousand times smaller than it is today. Agitation speeds within the starry 'gas' will approach the speed of light and from now on space will be intolerably hot. The cosmic background radiation will have reached a temperature of 3000 K, comparable with the surface temperature of the stars. Space, dark until this moment, will begin to redden rather like a piece of iron in the fire. Each photon in the cosmic radiation will have enough energy to shatter atoms, breaking the electromagnetic bonds between electrons and atomic nuclei.

Main events in the future of a closed universe			
After the Big Bang	**Before the Big Crunch**	**Absolute temperature (K)**	**Events**
15 billion years	−39 975 billion years	3	● Today
20 000 billion years	−20 000 billion years	0.03	● Maximal expansion ● Galactic emission blueshifted
39 999 billion years	−1 billion years	30	● Clusters of galaxies merge
	−70 million years	300	● Galaxies merge ● Space as hot as Earth environment today
	−600 thousand years	3000	● Atoms dissociate ● Universe red and opaque to photons
	−3 weeks	10^6	● Stars and planets destroyed by nuclear explosions ● Black holes merge
	−2 minutes	10^9	● Atomic nuclei disintegrate into protons and neutrons
40 000 billion years		?	● **(?) BIG CRUNCH**

FIGURE 4.3   Main events in the future evolution of a closed Universe. Time scales correspond to a cosmic density 1% greater than the critical value.

Henceforth, atoms in planetary and stellar atmospheres, as well as those in the interstellar medium, will begin to dissociate, freeing their train of electrons. As the temperature rises still further, cosmic radiation will begin to dissolve stellar surfaces and compress the stars, modifying their internal structure, whilst space itself will change colour from red through yellow, white and blue, becoming ever brighter.

Changes now take place faster and faster. Just a million years further on, the temperature will reach 10 million degrees, similar to the temperatures reigning in stellar cores. In an inferno of nuclear explosions, stars and planets will dissolve into a chaotic magma of nuclei, electrons and photons. Feeding upon this substance, black holes left over from the long-forgotten evolution of massive stars will grow and merge together to swallow up ever more of the Universe.

Three weeks later, the temperature will have risen to two billion K and photons in the cosmic background will have enough energy to break nuclear bonds between neutrons and protons in atomic nuclei. For a few minutes, the Universe will then enter an era equivalent to the one just before primordial nucleosynthesis occurred.

It would be difficult not to notice the extraordinary resemblance between the ideas described in the past few paragraphs, based on modern cosmology, and a passage in the famous *Eureka* by the American poet Edgar A. Poe. He had always shown a great interest in the physical sciences. On 3 February 1848, he gave a lecture in New York on 'The Cosmogony of the Universe'. The text of this lecture, published in *Eureka* the same year, was defined by the author as an essay on 'the Material and Spiritual Universe, its Essence, its Origin, its Creation, its Present Condition and its Destiny'. In this text Poe creates his own cosmology based on the meagre astronomical knowledge of his day, in the hope that 'it will (in good time) revolutionize the world of Physical and Metaphysical Science'. He never achieved his ambitious target because, in the words of American astrophysicist Edward Harrison: '… its science was too metaphysical and its metaphysics too scientific for contemporary tastes'. Poe was nevertheless the first to conceive of a dynamic Universe in global evolution, something which even Einstein did not manage to do three-quarters of a century later! Poe's Universe evolved under the influences of two opposing forces: a repulsion which shattered the unity of the primeval particle, and a Newtonian gravitational attraction tending to restore that unity. These ideas, which can be found at least in part in the writings of the pre-Socratic philosophers, were exposed in Eureka with a quite exceptional style, an extraordinary mixture of prose and poetry. Here is the passage mentioned earlier:

> … and the general result of this precipitation must be the gathering of the myriad now-existing stars of the firmament into an almost infinitely less number of almost infinitely superior spheres. In being immeasurably fewer, the worlds of that day will be immeasurably greater than our own. Then, indeed, amid unfathomable abysses, will be glaring unimaginable suns. But all this will be merely a climactic magnificence foreboding the great End. Of this End the new genesis described can be but a very partial postponement. While undergoing consolidation, the clusters themselves, with a speed prodigiously

accumulative, have been rushing towards their own general centre – and now, with a million-fold electric velocity, commensurate only with their material grandeur and with their spiritual passion for oneness, the majestic remnants of the tribe of Stars flash, at length, into a common embrace. The inevitable catastrophe is at hand.

Was it intuition or premonition perhaps? It is difficult to qualify this prophetic passage in any other way, especially when we realise that the solution of the cosmological equations was only obtained by Friedmann in 1922, and that a further twelve years went by before the British physicists Milne and McCrea showed that a similar solution (although approximate) exists in the context of Newtonian physics, the only kind of physics Poe could have known about at the time.

Swan or phoenix?

The appearance of the Universe during the final moments before the presumed Big Crunch will bear a remarkable resemblance to its earliest beginnings. The film will run backwards and our tale will end when the temperature reaches 10^{32} K, for beyond this so-called Planck limit, any attempt at a description based on today's physics will become impossible.

It is indeed a pity that Poe did not imagine the future of the human species in the blaze of its collapsing Universe. In Freeman Dyson's words:

> There is a great melancholy in the picture of a finite universe, its life force spent, its days of passion over, counting the hours remaining before it slides into oblivion. What will our last poets sing, whoever they may be, human or alien, as they watch the stars crowding together and streaming faster and faster across the imploding sky?

Guessing the answer to Dyson's question, we may affirm today that during the last million years before the Big Crunch matter will return to the same state of thermodynamic equilibrium it found itself in during the first million years following the Big Bang. But this time the gravitational entropy will be much higher than it was after the Big Bang. Most of the matter will now be in condensed form, swallowed up by black holes. For the first time in its history, the Universe will have attained maximal possible entropy. Hence the heat death occurs well into the future of the closed Universe, but in a quite different way to the one envisaged by nineteenth-century physics. Far from freezing up, the Universe will have an extremely high temperature. Unfortunately this temperature will be absolutely uniform, with no local variations, making energy flow impossible.

It thus seems that there really is no very-long-term hope for any life form in a closed Universe. Even someone like Dyson admits defeat, at least to begin with: 'No matter how deep we burrow into the earth to shield ourselves from the ever-increasing fury of the blue-shifted background radiation, we can only postpone by a few million years our miserable end.' But his optimism finally wins through, guided by an almost limitless confidence in the abilities of the human being. He thus imagines that: '... converting matter into radiation and causing energy to flow purposefully on a cosmic scale, we could break open a closed universe and change the topology of space–time so that only a part of it would collapse, and another part of it would expand forever.' It is difficult to know today whether such a solution could be found, even theoretically speaking. In fact, within the context of general relativity, it would appear to be impossible. However, taking quantum effects into account, the situation may be different. We can nevertheless admire the audacity of the proposal. It is hard to imagine a more grandiose project of cosmic engineering.

Would this be the swan song of the Universe? Or might the Cosmos be reborn somehow from its ashes, like the legendary phoenix, to begin a new cycle of expansion? Clearly, the prospect of an oscillating Universe, passing through an infinite sequence of expansion and contraction cycles, is extremely attractive from the philosophical viewpoint, as we saw at the beginning of this chapter. It should be noted, however, that the idea of Eternal Return is not necessarily to everyone's taste. Arthur Eddington made it clear that he was no great admirer of the phoenix. In 1928, he declared in his book *The Nature of the Physical World*: 'I would feel more content that the universe should accomplish some great scheme of evolution and, having achieved whatever may be achieved, lapse back into chaotic changelessness, than its purpose should be banalised by continual repetition. I am an Evolutionist, not a Multiplicationist. It seems rather stupid to keep doing the same thing over and over again.'

The modern cosmological version of the Eternal Return goes back to the work of Alexander Friedmann in 1922. He showed that at the end of each cycle the radius of the Universe goes to zero, putting an end to any physical continuity. In other words, the universes are completely separated, and there is no possibility of contact between them. About ten years later, American physicist Richard Tolman suggested that the zero radius singularity could be avoided in a realistic model. Contraction would proceed down to some minimal radius before transforming into a new expansion of the same Universe. Tolman observed that in this case the entropy of each cycle would be inherited from the previous one, since it could not be destroyed in the transition. The logical conclusion is that our own cycle could only have been preceded by a finite number of earlier cycles, at most one hundred, otherwise the present entropy would be much too high.

Tolman's idea was accepted for a long time, even though it only displaced the problem of the origin of the Universe, leaving open the question of how the first cycle could have emerged. However, it was not based on firm foundations, as the English physicists Stephen Hawking and Roger Penrose showed in the 1960s. The theorem which carries their name states that, according to general relativity, the singularity is inevitable, both in the past and in the future. Classical physics seems therefore to confirm Friedmann's view that there will be no resurrection for the Universe. But all hope is not lost for Tolman's hypothesis, because quantum gravity may offer a way round the initial and final singularities. It may even be that there is a solution to the problem of the continuously growing entropy from one cycle to the next. According to a recent suggestion, most of the entropy would be swallowed up by black holes just before maximal collapse. The new cycle would then begin with low entropy.

Among modern science fiction writers, one at least has dared to face the future of life in a closed Universe. In his book *Macrolife*, published in 1980, George Zembrowski imagines the very-long-term future of humanity, then of the galactic civilisation and finally of cosmic intelligence, in a Universe condemned to collapse upon itself. His prose does not have Olaf Stapledon's visionary power, although it is clearly inspired by him. But he is in tune with the developments of modern physics, as demonstrated by his suggestion that certain properties of rotating black holes might be used to escape from the final collapse, thereby allowing safe passage into a new cycle of expansion. Properties of these objects will also be put to use in one of the following sections. For the moment it is enough to know that, according to mathematical models, a rotating black hole is surrounded by a region called the 'ergosphere', in which space itself is dragged around by the rotation. No object could remain motionless inside this region, being rather like a small boat caught up in a whirlpool. On the other hand, with sufficient energy, someone could stay in the ergosphere for a considerable time without being swallowed up by the black hole. Here then is Zembrowski's plan, told by one of the representatives of the cosmic intelligence in this distant future:

> We will be circling within the ergosphere of that giant collapsar; and from there, we might be able to pass into the next cycle of nature, by moving through the neutral area near the singularity's equator… where centrifugal forces balance the crushing effects of gravitational collapse … the contraction will not continue to infinite density; long before the universe becomes very small, quantum effects will come into play, preventing ultimate collapse … expansion into new space will begin, carrying us with it, behind the white hole outstream.

Despite its strange properties, the ergosphere of the black hole is unlikely either to provide shelter from, or even to resist the growing fury of the contracting Universe. But you never know!

The slow decay of the open Universe

> Somewhere in the far north, in a nameless country, there is a huge, cube-shaped rock. Each side is a hundred kilometres long. Once every thousand years a small bird flies to the top of the rock and rubs its beak against it for a few moments. When the rock has disappeared, completely eroded by the rubbing of the bird's beak, nothing but a single day of eternity will have gone by.

Of course, this rock in Nordic mythology does not exist, and indeed it could not exist on our planet. With such dimensions, it would be more like a large asteroid. Let us suppose that the bird removes just one thousandth of a gram on each visit. It would then take about 10^{27} years to wear the rock away. This is 10^{17} times longer than the present age of the Universe and 10^{14} times longer than the duration of the closed Universe we investigated earlier. But the legend was clear, '… nothing but a single day of eternity ...' Let us take it literally. Our whole life of about hundred years contains roughly 35 000 days. A being whose days lasted as long as those in the legend would live for 10^{32} years.

However, even this enormous period is not eternity. The open Universe, infinite in space and time, provides a much vaster stage for the evolution of matter, as we shall see shortly. In the immensity of infinite space–time, the tens of billions of years during which stellar and galactic activity take place represent much less than the 10^{-43} seconds of the Planck time represent for us.

The trip to the final future of the open Universe was first undertaken at the end of the 1970s in three long articles published respectively by physicists Javal Islam, John Barrow and Frank Tipler, together with none other than Freeman Dyson once again. They showed that, in contrast with the past history of the Universe, events in the distant future do not contribute to the development of complexity. Very-long-term effects tend to dissociate all structures that the three forces (nuclear, electromagnetic and gravitational) have been able to create up until now, from nuclei and atoms to planetary, stellar and galactic systems. In such conditions, prolongation of life becomes problematic, as we shall see at the end of the present chapter.

It should be noted that recent evidence (obtained in 1998) suggests that our Universe may be behaving in an unexpected way. Observations of bright supernovae in remote galaxies seem to indicate that the expansion of the Universe is accelerating, contrary to the Standard Big Bang picture. This could be due to a peculiar form of energy, pervading the vacuum of space and dominating the dynamics of the Universe at large. In that case, all objects beyond the local supercluster of galaxies will recede with ever-increasing velocities from our own Galaxy. Their distances will increase exponentially with time and their apparent brightness will decrease in consequence. In about 150 billion years from now (ten times the present age of the Universe), even the closest of those galaxies will completely fade away. Our descendants, finding themselves in a Universe inhabited only by the few thousand galaxies of the local supercluster, will feel very lonely indeed!

However, this gloomy picture is still a long way from receiving decisive confirmation. We shall therefore assume that the standard picture of a Universe in decelerating expansion still holds. Following in the footsteps of these first explorers of the extreme future, we begin our voyage at the date 10^{13} years (10 000 billion years). By this time, we may be quite certain that all stellar activity will have ceased. Galaxies will lie at distances hundreds of times further than they do at present, whilst the temperature of the cosmic background radiation will have fallen to just 0.03 K. At this juncture, a typical galaxy will be composed of stellar corpses (black dwarfs, neutron stars and black holes), failed stars (brown dwarfs) and cold matter (planets, comets, asteroids and interstellar dust). These dark objects will still be linked to their parent galaxies via gravitational attraction. No light will be emitted from this point on, apart from an occasional short-lived glow as two dead stars collide somewhere near the galactic centre, where the density of stars is higher.

From time to time, a new star will emerge miraculously from such a collision and shine for several trillion years. This is simply because the brown dwarfs, stars with masses less than 0.08 times the solar mass, conserve all their hydrogen reserves intact. A collision between two brown dwarfs will occasionally produce a star whose mass exceeds this limit. The resulting red dwarf would then be able to burn its hydrogen. According to estimates by the American physicists Fred Adams and Greg Laughlin, a few hundred stars might be formed by this unusual process, occupying our Galaxy long after normal stellar activity has ceased. Moreover, a collision between two black dwarfs would give rise to a spectacular supernova explosion of a particular type. Referred to as type Ia supernovas (or SNIa), these explosions are the origin of most of the iron in the Universe, whilst oxygen and other heavy elements come from explosions of massive stars (type II supernovas, or SNII). In this way, iron production will continue over hundreds of trillions of years into the future, although at an extremely low rate.

It is no easy matter to calculate the detailed dynamical evolution of such a system in the long run. It depends on multiple gravitational interactions between the myriad bodies present. Each object feels the attraction of all the others, since gravity is a force with infinite range. It appears that an important role will be played by triple encounters in which the orbits and energy of a binary system will be perturbed by the approach of a third body. As a result of this cosmic billiards, some bodies will acquire sufficient energy and velocity to escape from the gravitational attraction of their partners. Some may even escape from

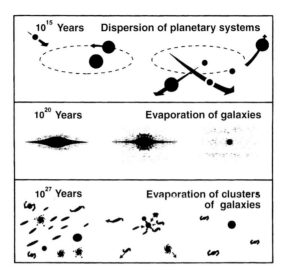

FIGURE 4.4 Long-term evolution of dynamical systems. Three-body gravitational interactions will gradually break up planetary systems, throwing planets from their orbits (top). The same phenomenon will lead to the evaporation of the greater part of the stars in a galaxy, whilst those remaining will condense into a galactic black hole with the mass of a billion suns (centre). On an even longer time scale, most galaxies will be ejected from their clusters, only a small proportion condensing to form supergalactic black holes (bottom).

their galaxy by the same process. Although these events are extremely rare, after around 10^{15} years (a million billion years) they will have ejected the great majority of planets from their orbits. Each planet will then follow its own trajectory within the galaxy, far from its original system. Some will even be ejected outside the galaxy and lost forever in the vast intergalactic space.

In a still longer term, after about 10^{19} years, the same process will have ejected most stars, or rather stellar corpses, from their parent galaxies. The small proportion remaining, fewer than 10%, will be the great losers in the game of galactic energy exchange. Stars faithful to their parent galaxy will continue along orbits that shrink further and further in towards the galactic centre. In this way, they lose even more energy in the form of gravitational waves, deformations of space–time caused by rotating objects, according to the theory of general relativity. Such energy losses are extremely low in normal conditions, but over these very long periods of time, they will end up sapping all the rest of the energy in the system and it will dwindle in size. Ultimately, it will collapse into a giant black hole, formed by the several billion surviving stars as they pile into a volume no greater than the Solar System. This will happen on a timescale of 10^{20} years, leaving a galactic black hole of mass a few billion solar masses. Over the whole of this period, the general appearance of the galaxy will be dominated by the contrast between a more and more condensed central part, eventually becoming a black hole, and a more and more dilute outer part destined to completely evaporate.

The same process will also act on the scale of clusters containing hundreds of thousands of galaxies. Most of their members will be expelled to suffer the fate already described above: evaporation of most of their stars and condensation of the remainder into a galactic black hole. By the end of about 10^{27} years, the other galaxies in the cluster will have collapsed to the centre of the system and formed supergalactic black holes with masses on a par with a thousand billion suns.

In the dark mists of time

Let us try to visualise the Universe in 10^{27} years from now. It will certainly be a much colder place. The temperature will be just 10^{-12} K. The wavelength of each photon in the cosmic background radiation will then be comparable with the present distance between Earth and the Moon. But the main feature will be a total absence of those majestic structures, planetary systems, galaxies and clusters of galaxies, which once gave the Universe its characteristic appearance and allowed us to get our bearings. From this time on, planets and asteroids, stellar and galactic corpses will follow their lonely path across perfectly black space, moving ever further apart as expansion proceeds.

It is hard to imagine the fate of any civilisation in this distant future. Assuming that one did exist, could it somehow prevent the slow decay of the parent galaxy? By mobilising its resources, it might be able to organise stellar encounters in such a way that those stars likely to make an escape would remain bound to their system. A domino technique would be required. For example, a comet could be used to perturb the trajectory of a planet, which in turn would modify the orbit of a star, and so on. Such a project would necessitate extraordinarily complex and precise manoeuvres, although doubtless within the capabilities of future astro-engineers. It would be important to begin early on and make a very-long-term investment, since the fruits of these operations would only be forthcoming hundreds of billions of years down the line. A type III civilisation might thereby delay the evaporation of the stars from its galaxy by some 10^{23} years. However, this suggestion of Barrow and Tipler does not take into account the extreme energy shortage prevailing in this distant epoch. There would be no more active stars, nor even any gas clouds able to fuel nuclear fusion. The main preoccupation of any survivors would be to prevent their planet from flying off alone into intergalactic space, by keeping it in an orbit around a galactic black hole. For strange though it may seem, these space whirlpools provide a faint hope of survival.

When a black hole forms, it may well not be static, but rather in a state of rapid rotation about its axis. In fact, if the initial stellar or galactic system begins with a slight rotation, collapse will amplify it, just as ice skaters can increase their speed of rotation by bringing in their outstretched arms and holding them against the body. The properties of rotating black holes were studied in the 1960s and it was shown in particular that their rotational energy can reach quite significant levels, up to 30% of their mass energy. As a comparison, nuclear reactions transform only a few thousandths of the masses involved into energy.

According to the theory, these objects are surrounded by a region in which space–time itself is forced to take part in the rotation, dragged around by the motion of the black hole. An object can escape from this region if it has enough energy. The American physicist John A. Wheeler called it the ergosphere, from the Greek word *ergon* meaning 'work'. It plays a key role in the energy extraction process imagined by Roger Penrose. A projectile is fired into the ergosphere and breaks into two parts. One is swallowed up by the black hole and the other is ejected. Penrose showed that if the fragment finally captured follows a retrograde orbit (moving in the opposite direction to the rotation of the black hole), the fragment that comes back out has greater energy than the initial energy of the projectile. Naturally, the bonus energy gained by the ejected fragment is lost by the black hole, which consequently loses rotational speed.

We may thus imagine future civilisations locating themselves in the vicinity of stellar or galactic black holes, using their rotational energy and getting rid of their rubbish at the same time. It would be hard to think of a more ecological solution to this problem: any waste in the Universe simply disappears into the monster's mouth. But the energy reserves of the rotating black hole are not infinite. A type II civilisation, consuming energy at a rate comparable to that radiated by the Sun today, could survive another 10^{23} years by exploiting a galactic black hole in this way (always assuming there is enough material in the neighbourhood to feed the giant). Even the rotation of the biggest supergalactic black holes, with mass 10^{14} solar masses, will be brought completely to a halt beyond about 10^{27} years at this rate of energy consumption. Of course, by drastically reducing the rate and moving down to level I civilisation, survival could be extended as far as about 10^{35} years. But even these sacrifices are not enough to endure for the whole of eternity.

Protons are not forever

The various metamorphoses of the open Universe described in the last section are due to macroscopic effects of gravity. Although once at the origin of cosmic order, this force will finally destroy all the bound systems it ever created, either by hurling their components into black holes or by expelling them into intergalactic space. In the longer term, it is microscopic effects that threaten the existence of individual objects, and even the very survival of inert matter.

Every material object is basically composed of the fundamental particles protons and electrons. These particles were created during the first moments of the Big Bang. Under the influence of nuclear, electromagnetic and gravitational forces, they subsequently formed nuclei, atoms and molecules, which make up all material structures we know today. These formations are mortal in the sense that they can be destroyed by the action of those same forces. However, their basic constituents are not supposed to disappear, merely to separate, remaining available for recombination into other nuclei, atoms and molecules at a later stage. This traditional view of the structure of matter can in fact be traced back to Democritus and other Greek Atomists. But over the past thirty years, these ideas have been put into question by modern physics.

According to Grand Unification Theories (GUTs), which attempt to describe all the forces of nature (apart from gravity) within a single coherent formalism, the proton itself should actually be unstable. Just like a radioactive particle, it should decay into other lighter particles: neutrinos, photons, electrons and their antiparticles, positrons. Fortunately for us, this decay takes place at an exceptionally slow rate, requiring at least 10^{31} years according to the simplest model.

It might be thought that this very long period, 10^{21} times longer than the present age of the Universe, would make any kind of check quite impossible. However, the statistical nature of the process actually makes it observable because it implies that 1 proton out of every 10^{31} should decay each year. The idea then is to observe a sufficiently large amount of matter, let us say 1000 tonnes of water, which contains around 10^{32} protons, for a period of one year, and hope that a handful of decays will be detected. Many experiments of this kind were carried out in the 1980s, but without success. Their failure weakens the theory, at least in its simplest form. But it does not necessarily mean that the proton is stable. It may just be that its lifetime is (much) longer than 10^{31} years. In the absence of any definitive result concerning stability or instability, we must treat each of the two possibilities on an equal footing.

Let us assume, for example, that the proton lifetime is 10^{32} years, that is, we expect one proton out of every 10^{32} to decay every year. A black dwarf or a neutron star would then witness 10^{25} proton decays each year, these protons being replaced by electrons, positrons, neutrinos and photons. Apart from the neutrinos, which interact only very weakly with matter and which would escape immediately, the other particles would be absorbed into the star and their energy would heat it up. Calculations show that the star's temperature would depend on its density: a few degrees for a black dwarf and a hundred degrees or so for a neutron star. Hence, even though their own energy reserves were long since depleted, these stars would continue to radiate weakly, even after 10^{14} years. The power radiated by the surface of a black dwarf would amount to a few hundred watts, roughly the same as an electric light bulb. To all intents and purposes, these objects would remain invisible. Moreover, energetic photons from proton decays would break up atomic nuclei inside the black dwarf (essentially carbon and oxygen nuclei), forming other, lighter ones, such as helium. The chemical composition of these objects would thus be modified as time went by. They should have completely evaporated by the end of 10^{34} years at the very most. At this point, all the protons in nature, including those in the interstellar medium, will have decayed. Of course, if the proton lifetime is much greater than 10^{32} years, the evaporation process will take longer and the corresponding radiation will be much weaker.

At this distant epoch, the Universe will contain only a rarefied gas of electrons, positrons, neutrinos and photons, with the occasional black hole scattered here and there. The gas will have an extremely low density. The average distance between an electron and a positron will be several light-years (similar to the distance between ourselves and the nearest star α Centauri). The probability of their meeting and annihilating one another will be virtually zero. These two types of particle, more massive than the other survivors, will then dominate the cosmic plasma for all eternity. As far as the black holes are concerned, these last remaining massive objects will experience absolute solitude, their nearest neighbours lying at 10^{19} light-years from them, a billion times further than the most distant galaxies that we observe today.

Black holes die too

Whether the proton is stable or not, another quantum effect will not fail to occur in the long run: evaporation of black holes. In classical physics, a black hole can only grow, by absorbing matter and energy which it then lays up for all eternity. However, in 1974, Stephen Hawking showed that a black hole is not a chasm of no return. In fact, in certain conditions, it can emit thermal radiation. Hawking reached this extraordinary conclusion by bringing together two theories quite foreign to one another up until this time: general relativity (the macroscopic theory of gravity) and quantum mechanics (the theory of microscopic phenomena). Even though this union has not yet come to fruition in a wider context, Hawking's conclusions about radiation from black holes are generally accepted in the physics community.

At the heart of these effects are the strange properties of the quantum vacuum. We have already observed that the Universe itself may have sprung from a fluctuation in this ethereal substrate in the extraordinary conditions prevailing during the Planck period. In more moderate conditions, fluctuations in this 'vacuum' simply create particle–antiparticle pairs. Such pairs are said to be virtual since they immediately annihilate one another, having survived only a tiny fraction of a second. (Particles are always created in pairs so as to conserve the zero electric charge of the fluctuation.) However, if the fluctuation occurs near the horizon (the imaginary 'surface') of a black hole, the resulting pair may be separated. One of the particles then falls into the black hole, whilst the other manages to escape. An external observer would only witness the exiting particle (electron, positron, neutrino or photon) and would naturally conclude that the black hole is radiating.

Hawking showed that, because of this particle and photon emission, the black hole can be treated as a hot body with a characteristic temperature inversely proportional to its mass. It transpires that for stellar black holes the temperature is extremely low, of order 10^{-8} K. For galactic black holes it is even lower, something like 10^{-17} K, because they are much more massive, weighing in at around a billion solar masses. These temperatures are currently a long way below the cosmic background temperature, and they will remain so for several billion billion years. Consequently, black holes will long continue to play their role of cold reservoir, absorbing more heat from their environment than they will give up to it.

After some 10^{30} years, however, the temperature of the Universe will have fallen below these values. First the stellar black holes, followed by the galactic and eventually the supergalactic black holes, will then be hotter than their environment. They will begin to emit their characteristic radiation, at the expense of their mass, which will gradually be converted into radiation. As its mass decreases, the temperature of a black hole will rise steadily, leading to more and more intense radiation, and eventually, the total evaporation of the object. Slightly before its final disappearance, with a temperature above a few billion degrees, evaporation will become explosive. After a long period of absolute darkness, the shadows of this remote future will be momentarily banished. A stellar black hole will take 10^{66} years to evaporate completely, and a galactic or supergalactic black hole even longer, more like 10^{100} years.

Energy emitted by proton decay, if it occurs, and evaporation of black holes will dominate the extraordinarily dilute and cold cosmic

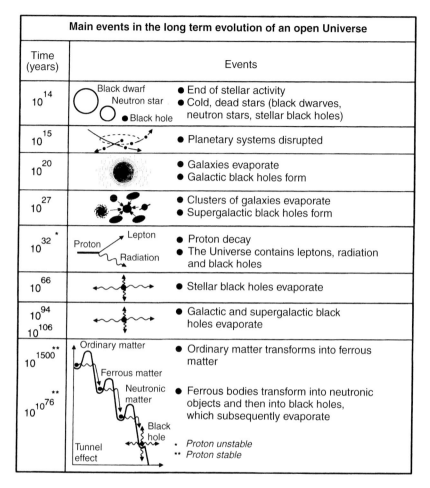

Main events in the long term evolution of an open Universe	
Time (years)	Events
10^{14}	● End of stellar activity ● Cold, dead stars (black dwarves, neutron stars, stellar black holes)
10^{15}	● Planetary systems disrupted
10^{20}	● Galaxies evaporate ● Galactic black holes form
10^{27}	● Clusters of galaxies evaporate ● Supergalactic black holes form
10^{32*}	● Proton decay ● The Universe contains leptons, radiation and black holes
10^{66}	● Stellar black holes evaporate
10^{94} 10^{106}	● Galactic and supergalactic black holes evaporate
10^{1500**} $10^{10^{76}**}$	● Ordinary matter transforms into ferrous matter ● Ferrous bodies transform into neutronic objects and then into black holes, which subsequently evaporate * *Proton unstable* ** *Proton stable*

FIGURE 4.5   Main events in the long-term evolution of an open Universe.

background radiation at such remote epochs. In fact, evaporating black holes might be able to supply the needs of any surviving civilisations until they completely disappear. It is a strange fate indeed for these singular objects that they should become the last energy source and the only hope of survival in the dark night of the Universe. Their radiation will last for so long that the 10^{30} past years of the Universe will look even shorter than the Planck time does for us today!

Assuming the proton to be stable, future civilisations could thus extend their life span up to around 10^{100} years. They could even survive beyond this date using the fact that a black hole's evaporation period is proportional to the cube of its mass. A ten-fold mass increase means a thousand-fold increase in the time taken for evaporation. Playing a game of cosmic billiards, future engineers might bring black holes together, increasing their mass and slowing the emission rate, according to the American physicist Stephen Frautschi. They would have to wander further and further afield in search of black holes and projectiles (asteroids, planets, dead stars, and so on) suitable for their cosmic game. But the desperate struggle to ward off the inexorable decay of the Cosmos could not succeed forever.

Unstable future

If the proton is unstable with a lifetime shorter than 10^{66} years, no macroscopic object will exist in the Universe during the long period of evaporation of the black holes. Only the extremely dilute gas of light particles, such as electrons, positrons, neutrinos and the inevitable photons of the cosmic background radiation, will fill out the vast extent of the Universe. However, theories of microphysics suggest that the proton might have a much longer lifetime, up to 10^{100} or 10^{200} years, which would be impossible to check experimentally. If this were true, the phenomena related to proton decay (described in a previous section) would take place much more slowly, right through the long period of black hole evaporation. Finally, if the proton is stable, macroscopic objects such as asteroids, planets and dead stars, which have escaped from dynamical evaporation of the galaxies, will also populate the Universe in 10^{100} years from now. The mean distance separating any two of these objects will then be of order 10^{45} light-years, 10^{35} times greater than the present size of the observable Universe.

It seems that no energy reserves will remain beyond this point. And yet, tiny changes will continue to take place on the microscopic scale, quite imperceptible even during the black hole evaporation. Although extremely slow, these phenomena will eventually have important consequences, resulting in the decay of every massive body.

At the root of this slow decay is a well known phenomenon of quantum mechanics called the tunnel effect. A classical particle has no chance of ever passing to the other side of a wall if it does not have the energy needed to jump over. A quantum particle, on the other hand, has a non-zero probability of succeeding. Tests of this phenomenon are commonplace in the laboratory. It allows the systems we are now considering to follow a general tendency of any physical system: to evolve towards ever more stable states of lower energy. Of course, excess energy is radiated into space during the process and the entropy of the Universe continues to increase.

Among the atomic nuclei making up the various material bodies, the most stable is iron-56, with 26 protons and 30 neutrons, king of all nuclear creation. Its maximal stability represents a kind of ideal for all the other nuclei, which try to reach it through nuclear reactions. Lighter nuclei proceed through fusion and heavier ones through fission, increasing and decreasing their masses, respectively. In the context of classical physics, the system requires an external energy intake in order to trigger such reactions. For example, if nuclear fusion is to occur in stellar cores, the temperature must reach tens or hundreds of millions of degrees. As we have seen, this is brought about through the action of the gravitational force. However, even in the absence of this activation energy, there is nevertheless an infinitesimal probability that the reactions will take place by the tunnel effect. This happens at an extremely slow rate on our time scale. But nothing is slow when you have all eternity to do it! The time required for black dwarves, planets, asteroids and dust to transform into iron is estimated at 10^{1500} years.

From a nuclear point of view, the ferrous state of matter is certainly the most stable. But there exist even lower energy states involving no atomic nuclei at all. Hence, within the ferrous objects, iron-56 nuclei will be slowly metamorphosing. Their protons combine with nearby electrons, which neutralise their electric charge and turn them into neutrons. In fact, a crystal lattice of neutrons has even lower energy than the iron state, and is more stable. After a much longer period than the last one mentioned, all solid objects in the Universe will reach the same neutronic state.

Even the quite impeccable organisation of these neutron crystal lattices does not manifest the ultimate level of stability. Through the tunnel effect, a particle at the surface of a neutronic object has a non-zero probability of either acquiring the speed necessary to escape from the body or else slipping towards the interior. The body thus gradually loses mass from its outer regions, whilst the remainder condenses further and further to end up as a black hole. We have already encountered a similar situation produced by the dynamical evolution of self-gravitating systems. Most galaxies evaporate off, the remainder collapsing down to produce a supermassive black hole. The time scales in the two cases differ enormously, however. In the case of self-gravitating systems, it lies somewhere between 10^{15} and 10^{20} years. For tunnelling evolution of sub-stellar mass objects to black holes, Dyson has estimated the time as $10^{10^{76}}$ years. This may well be the largest number ever mentioned in physics and is certainly beyond human comprehension!

The last stage in the future evolution of matter will be the evaporation of these newly born black holes via Hawking's mechanism. This will take much less than 10^{66} years, a brief instant compared with the $10^{10^{76}}$ years required to produce them by the tunnel effect. According to all present evidence, the rest of eternity will belong to electrons, positrons, neutrinos and photons, the only entities capable of existing through the endless night of the expanding Universe.

Impossible eternity?

What shall I do or write
against the fall of night?

A. E. Housman, *Transcience*

And so our voyage to the edge of the future has come to an end. This is as far as we can predict on the basis of our present understanding of the physical world. Modern cosmology has shown us a far more complex future than was predicted by nineteenth-century physics, much richer in events and lasting much longer. However, even though the entropy of the Universe never attains its maximal value, the final result of this long haul is a picture that hardly differs from heat death. Almost all matter will be converted into dilute and cold radiation. A few light particles will be dispersed across the immensity of a Universe in continual expansion. Faced with such a prediction, it is hard not to be pessimistic about the fate of intelligence, so well expressed in Bertrand Russell's famous passage:

... that all the labours of the ages, all the devotion, all the inspiration, all the noonday brightness of human genius, are destined to extinction ... and the whole temple of Man's achievement must inevitably be buried beneath the debris of a universe in ruins – all these things, if not quite beyond dispute, are yet so nearly certain that no philosophy which rejects them can hope to stand.

But this dark destiny did not overly perturb him.

I am told that that sort of view is depressing and that if people believed it, they would not be able to go on living ... But nobody really worries about what is going to happen millions of years hence. Therefore, although it is of course a gloomy view to suppose that life will die out ..., it is not such as to render life miserable. It merely makes you turn your attention to other things.

Russell's philosophical positivism is clear. But others find it much harder to accept the philosophical implications of the disappearance of life. In his *Autobiography*, Charles Darwin expresses his consternation in the face of evolution's failure: 'Believing as I do that man in the distant future will be a far more perfect creature than he now is, it is an intolerable thought that he and all other sentient beings are doomed to complete annihilation after such long-continued slow progress.'

These feelings are entirely shared by Herbert G. Wells, author of *The Time Machine*. In 1902, during a conference on the *Exploration of the Future* at the Royal Institute in London, Wells expressed his faith in the future of the human species, whilst fully recognising that the evidence went against him:

And finally there is the reasonable certainty that this sun of ours must some day radiate itself toward extinction; ... that this earth of ours will be dead and frozen, and all that has lived upon it will be frozen out and done with. There surely man must end. That of all such nightmares is the most insistently convincing. And yet one doesn't believe it. At least I do not ... because I have come to believe in certain other things: in the coherency and purpose in the world and in the greatness of human destiny. Worlds may freeze and suns may perish, but there stirs something within us now that can never die again.

Other navigators of the remote future, such as Dyson, Frautschi, Barrow and Tipler, are evidently closer in spirit to Darwin and Wells than to Russell. Their concern has been to explore ways of prolonging the reign of life. It is difficult to overcome the basic problem of energy supply, at least within the limits of our present knowledge. Annihilation of electron–positron pairs and radiation by black holes are interesting possibilities. But they are nevertheless non-renewable resources. More and more remote supplies of electrons and positrons must therefore be sought for mutual destruction, or black holes for fruitful merger with their own kind. As time goes by the quest for cosmic fire can only get more difficult, and expansion will make the distances insurmountable. It turns out that expansion will only be slow enough to apply these techniques if the density of the Universe happens to be exactly equal to the closure density (the marginally closed Universe). In the more likely case of an open Universe, expansion will be too rapid and future civilisations will never be able to cross cosmic distances fast enough.

The only way out glimpsed by Barrow and Tipler is to take advantage of anisotropies in the Universe, if there are any. Indeed, up to this point we have considered the future in the context of the standard Big Bang model. In this view, the Universe is assumed to be perfectly isotropic, expanding at the same rate in every direction. This implies that cooling happens in the same way everywhere and that the cosmic background temperature at a given time is identical at all points. However, it may well be that reality does not conform to such a simple model. Our experience shows that even the most successful models are just an approximation to the real world. (For example, the planets and stars are often modelled by spheres, a simple and useful working hypothesis, whilst it is clear that no real body can be perfectly spherical.) Any anisotropy would reveal itself through more rapid expansion in some directions than in others, leading to greater cooling in those directions. Barrow and Tipler suggest using the temperature difference between these colder regions and neighbouring warmer regions as an energy source, although they do not go into the details.

However, the energy shortage will not be the only threat hanging over civilisations in the remote future. The instability of matter caused by proton decay or tunnel effect will pose equally serious difficulties, destroying every solid configuration in the long term. One particular question, although rather academic at the present time, will become all important: does intelligent existence depend on the material substance supporting it? In other words, does it depend on the organic molecules making up our brain, or on their underlying structure, that is, their disposition in space? Put another way, would it be possible to build an exact copy of the brain, able to function in the same way, but using other materials?

We do not know the answer to this question at the present time, although it promises to be crucial for the very-long-term future of intelligence. Clearly, if the answer is negative and intelligence always requires an organic basis, then it is condemned to disappear from the cosmic scene. The first to have approached this question in the affirmative seems to have been the British physicist Desmond Bernal. In his book *The World, the Flesh and the Devil*, mentioned in Chapter 2, Bernal attempted to explore the very-long-term future of our species. The instability of matter was unknown at the time, but the spectre of heat death for the Universe encouraged him to seek alternative future life forms:

> Finally, consciousness itself may end or vanish in a humanity that has become completely etherialized, losing the close-knit organism, becoming masses of atoms in space communicating by radiation, and ultimately perhaps resolving itself entirely into light... these beings, nuclearly resident, so to speak, in a relatively small set of mental units, each with the bare minimum of energy, connected together by a complex of etherial intercommunication, and spreading themselves over immense areas and periods of time, by means of inert sense organs which, like the field of their active operations, would be, in general, at great distance from themselves. As the scene of life would be more the cold emptiness of space than the warm, dense atmosphere of the planets, the advantage of containing no organic material at all, so as to be independent of these conditions, would be increasingly felt.

The visionary Bernal believed that even this strategy could not prolong life indefinitely. But he did believe that life could be made to last several million times longer by appropriate restructuring, before falling foul to the unavoidable heat death.

The idea of a dematerialised intelligence, freed from the constraints and comforts of its planetary cradle and living in the chilly liberty of space, was subsequently taken up by several science fiction writers. The best known of these is surely the black cloud, described in 1957 by British astrophysicist Fred Hoyle in his science fiction novel of the same name: *The Black Cloud*. A huge opaque cloud, with dimensions comparable to those of Earth's orbit, enters our Solar System and places itself between ourselves and the Sun. Because it absorbs the Sun's light, there is a sudden fall in temperature on Earth causing great destruction to life. The cloud turns out to be endowed with superior intelligence and scientists manage to communicate with it. They discover that it is a cloud of electrically charged particles, bound together and communicating with one another through electromagnetic forces. After several weeks stocking up with solar energy, the cloud eventually sets off again, and life on Earth can gradually recover.

The ideas originally expressed by Bernal, Hoyle and others can be more rigorously formulated today in the language of cybernetics. According to Barrow's and Tipler's views, exposed in their master work, *The Anthropic Cosmological Principle*, an intelligent being can be considered as a sort of computer. The way it works is governed by the laws of physics. Intelligence itself then corresponds to the program, or the software, whilst the human body is the hardware. Pushing the analogy to the extreme, Barrow and Tipler go as far as to identify the program which controls the body with the religious notion of the soul, since both are dematerialised entities. Without going into the details of their metaphysics, it is interesting to see that the question of the survival of intelligence is then reformulated as follows: Will it be possible in the remote future to build computers on which complex programs can run? Will there be any obstacles to their operation?

Freeman Dyson was the first to discuss these ideas seriously, in 1979, in his famous article so eloquently entitled *Time without End: Physics and Biology in an Open Universe*. In this article, which opened the way to scientific eschatology, Dyson proposes to transplant future intelligence into a cloud of charged particles such as electrons and positrons, since these are in principle eternal. The electromagnetic properties of these particles would guarantee the cloud's cohesion and also its internal and external communications by emission of radiation.

Such dematerialised beings would require a minimum of energy to maintain their structure in working order and keep their entropy at a low level, vital functions of a living being. Some energy would also be used to process information, an essential function of an intelligent being. In both cases, the minimal energy can only be greater than the energy of ambient photons. In other words, the temperature of the organism must be higher than the cosmic background temperature.

As the latter decreases with time, these ectoplasmic beings could adjust their operating temperature downwards, ever closer to ambient temperature, in such a way as to reduce their vital energy needs. Naturally, with less energy input, their metabolism would become slower, as would their reflexes, thinking and all other functions. This would not raise any particular problems because they would have all the time in the world to accomplish their projects! In Dyson's view, the amounts of energy required for survival in this distant future would be rather modest, all things considered. A civilisation with the same degree of complexity as our own could prolong its existence for all eternity using just the energy radiated by the Sun in eight hours. This surprising result is due to the extremely low level of the ambient temperature in the distant future. Various physical processes could therefore be carried out very slowly and with minimal energy consumption.

Dyson finds, however, that even a slowed-down metabolism could not guarantee eternal life for these beings, for the following reason. Like every living being, they must evacuate heat produced when they operate in order to keep their internal entropy at a low level. Of course, the rate of heat evacuation is proportional to the temperature difference between the organism and its environment. But this temperature difference must decrease as time goes by if the organism is to extend its life span. A time will come when, despite its slowed metabolism, heat can no longer be removed quickly enough and internal entropy will rise, leading to decay and death.

In order to avoid death by internal overheating in a cold Universe, Dyson suggests an intermittent type of life, operating over shorter and shorter periods, separated by longer and longer periods of hibernation. During these periods, the organism's vital functions would stop completely, whilst internal heat continued to be dissipated. By this strategy, Dyson finds that life and intelligence could survive for ever, even communicating across the vast reaches of an expanding Universe. We may contrast Russell's resigned pessimism with Dyson's basic optimism: 'No matter how far we go into the future, there will always be new things happening, new information coming in, new worlds to explore, a constantly expanding domain of life, consciousness and memory.'

Again, it is difficult to know whether Dyson's wild speculations have a physical basis or not. In fact, recent studies based on thermodynamical and quantum properties of computation, conclude that sentient material beings cannot live eternally in an ever-expanding Universe. With finite resources at their disposal, they will not be able to continue thinking forever, or even to switch between hibernation and consciousness.

So who could be right, Russell or Dyson? Only eternity will tell.

Conclusion

This is a work of fiction. I have tried to invent a story which may seem a possible, or at least not wholly impossible, account of the future of man ...

To romance of the future may seem to be indulgence in ungoverned speculation for the sake of the marvellous. Yet controlled imagination in this sphere can be a very valuable exercise for minds bewildered about the present and its potentialities. Today we should welcome, and even study, every serious attempt to envisage the future of our race; not merely in order to grasp the very diverse and often tragic possibilities that confront us, but also that we may familiarize ourselves with the certainty that many of our most cherished ideals would seem puerile to more developed minds. To romance of the far future, then, is to attempt to see the human race in its cosmic setting, and to mould our hearts to entertain new values.

But if such imaginative construction of possible futures is to be at all potent, our imagination must be strictly disciplined. We must endeavour not to go beyond the bounds of possibility set by the particular state of culture within which we live. The merely fantastic has only minor power. Not that we should seek actually to prophesy what will as a matter of fact occur; for in our present state such prophecy is certainly futile, save in the simplest matters. We are not to set up as historians attempting to look ahead instead of backwards. We can only select a certain thread out of the tangle of many equally valid possibilities. But we must select with a purpose. The activity that we are undertaking is not science, but art; and the effect that it should have on the reader is the effect that art should have.

Yet our aim is not merely to create aesthetically admirable fiction. We must achieve neither mere history, nor mere fiction, but myth. A true myth is one which, within the universe of a certain culture (living or dead), expresses richly, and often perhaps tragically, the highest aspirations possible within that culture. A false myth is one which either violently transgresses the limits of credibility set by its own cultural matrix, or expresses aspirations less developed than those of its culture's best vision. This book can no more claim to be true myth than true prophecy. But it is an essay in myth creation ...

If ever this book should happen to be discovered by some future individual, for instance by a member of the next generation sorting out the rubbish of his predecessors, it will certainly raise a smile; for very much is bound to happen of which no hint is yet discoverable. And indeed even in our generation circumstances may well change so unexpectedly and so radically that this book may very soon look ridiculous. But no matter. We of today must conceive our relation to the rest of the universe as best we can; and even if our images must seem fantastic to future men, they may none the less serve their purpose today.

So began Olaf Stapledon with Shakespearean style in his preface to *Last and First Men*, published in 1930. It would be difficult to find a better illustration of the purpose of this book.

In particular the passage shows how we should consider our visions of the future. This is not some distant land that we may hope one day to visit, out of simple curiosity, carried on the wings of modern science. It is better viewed as a screen on which we may project the higher aspirations of our culture. To this end, we must, of course, reflect upon those aspirations.

The decline of religion and philosophy, the unsurpassed myth creators of the past, has left the way open to modern science. Since the last century, science has become one of the main sources of modern myth in our society. This may seem paradoxical, since the scientific method is based on doubt and scepticism and is supposed to do away with myth. However, in its approach, science has regularly produced new ones. One of these is the myth of all-powerful humans who, through science, are able to raise themselves above nature and indeed to become masters of their own destiny. Right through this book, we have seen the modern version of this myth, suggesting that mankind might extend its empire right across the immensities of space and time.

For some critics the exaltation of space travel is nothing but a vast extension of occidental imperialism. The words used are not innocent, but reveal either conscious or unconscious motivations in those who promote 'colonisation' of the planets, 'exploitation' of their resources and 'conquest' of space.

Yet other critics see this glorification of our cosmic future merely as a way of running with the tide, providing a new system of beliefs to replace others in decline. The English philosopher Mary Midgley criticises these visions of our cosmic future particularly severely. She finds them not only useless but even dangerous. The eloquent conclusion to her book *Science as Salvation* concludes in controversial style:

> Our planet is in deep trouble; we had better concentrate on bailing it out. At this point, to keep up one's spirits by further orgies of self-congratulation may be a natural reaction, but it is a dead end. Paranoia, if further encouraged, is liable finally to undermine all wish to get back in touch with reality. The discrepancy between image and fact is growing too wide to be tolerated. For the general sanity, we need all the help we can get from our scientists in reaching a more realistic attitude to the physical world we live in.

I find these criticisms rather unfair and baseless. Whilst recognising the seriousness of our planet's state of health, I believe that futuristic visions do not prevent our species from solving its problems. It was not the dreams of Tsiolkovski or Wells that led to a loss of contact with reality. In fact, they have helped mankind to a better understanding of our place in the world, and they have encouraged us to set ambitious goals in order to improve it, and to excel in attaining them. In the process, we have redefined our relationship with the world. In other words, these visions help mankind to evolve culturally. In the long process of hominisation, the discovery of new frontiers and exploration of new territories have always played a basic role. The British philosopher Alfred North Whitehead wrote: 'Without adventure, civilization is in full decay.'

Today it is difficult to know whether space will play the role of the new frontier. If so, it is clearly not for the immediate future. Even though we may have the technological knowhow to take the first steps, I do not think that an ailing civilisation such as our own should throw itself into undertakings of this magnitude. Earth must first heal its wounds. But as time goes by, the cosmic expanses will beckon all the more urgently.

And nor can we know what our cosmic adventure will be. The difficulties will certainly be more formidable than we can imagine today. In particular, any encounter with an extraterrestrial civilisation would dramatically alter our approach. Some see a hope of salvation, and some a threat. Personally, I think it will be a lesson in humility. If on the other hand we find that we are the only form of intelligence in the Galaxy, it will be our duty to preserve nature's uniquely successful experiment and spread it across the whole Universe. In this case, as Herbert G. Wells wrote: 'Our choice is limited: either the whole Universe or nothing.'

BIBLIOGRAPHY

• General public
+ Technical
* Science fiction

GENERAL REFERENCES

*Aldiss, B. (1988) *Trillion Year Spree: A History of Science Fiction.* London: Palladin. A standard reference on the history of science fiction.

•Audouze, J. & Israel, G. (eds.) (1994) *The Cambridge Atlas of Astronomy.* Cambridge: Cambridge University Press. A magnificent and richly illustrated presentation of modern astronomy.

•Bernal, J. D. (1969) *The World, the Flesh and the Devil.* Indiana University Press. First published in 1927.

•Clarke, A. C. (1962) *Profiles of the Future.* London: Victor Gollancz.

•Midgley, M. (1992) *Science as Salvation.* London: Routledge. An interesting but sometimes exaggerated criticism of futuristic visions.

•Morisson, D. (1993) *Exploring Planetary Worlds.* Scientific American Book.

•Neal, V. (ed.) (1994) *Where Next, Columbus? The Future of Space Exploration.* Oxford: Oxford University Press.

•Sagan, C. (1995) *Pale Blue Dot: A Vision of the Human Future in Space.* London: Headline Book Publishing. Beautifully illustrated, Sagan's penultimate book presents a resolutely optimistic vision of our future in space.

•Sagan, C. & Shklofski, I. (1966) *Intelligent Life in the Universe.* Holden Day. A classic of its kind which has enchanted generations of readers. Probably the best book on this subject.

•Schneider, J. & Leger-Orine, M. (eds.) (1987) *Frontiers and Space Conquest.* Kluwer.

CHAPTER 1

+ Adelman, S. (1981) Can Venus be transformed into an Earth-like planet? *Journal of the British Interplanetary Society*, **35**, 3–8.

• Bacon, F. (1992) *New Atlantis*. Kila, Minnesota, USA: R.A. Kessinger Publishing Company.

• Beardsley, T. (1996) Science in the sky. *Scientific American*, June issue, 36–42.

• Berry, A. (1995) *The Next 500 Years*. London: Headline Book Publishing.

• Bonnet, R. (1993) L'Avenir des télescopes spatiaux. *Ciel et Espace*, December issue, 40–43.

• Booth, N. (1992) *Espace, les 100 prochaines années*. Larousse.

+ Burnes, J. (1995) Astronomy from the Moon. *Robotic Telescopes*, ASP Conference Series, **79**, 242–251.

* Clarke, A. C. (1992) *The Fountains of Paradise*. Ballantine Books.

• Clarke, A. C. (1994) *The Snows of Olympus*. London: Victor Gollancz. An excellent description of our view of Mars through the centuries, together with the various stages involved in terraforming the planet.

• Esterle, A. (ed.) (1993) *L'Homme dans l'Espace*. Paris: Presses Universitaires de France.

• Jean, G. (1994) *Voyages en Utopie*. Découvertes Gallimard.

• Lardier, C. (1992) *L'Astronomie soviétique*. Armand Colin.

• Lewis, J. (1996) *Mining the Sky*. New York: Addison Wesley.

+ Lewis, J., Matthews, M. & Guerrieri, M. (1995) *Resources of Near-Earth Space*. The University of Arizona Press. A mine of information concerning the resources of the Solar System and possible approaches to exploiting them.

+ McKay, C., Toon, O. & Kasting, J. (1991) Making Mars habitable. *Nature*, **352**, 489–496.

+ O'Neill, G. (1974) The colonisation of space. *Physics Today*, September issue, 32–44.

• O'Neill, G. (1977) *The High Frontier. Human Colonies in Space*. New York: William Morrow.

* Poe, E. A. (1987) The unparalleled adventure of one Hans Pfaal. In *The Science Fiction of Edgar Allan Poe*, p. 12, London: Penguin Books.

+ Pollack, J. & Sagan, C. (1995) Planetary engineering. In *Resources of Near-Earth Space*, eds. J. Lewis, M. Matthews & M. Guerrieri, pp. 921–950. The University of Arizona Press.

+ Rettig, T. & Hahn, J. (eds.) (1996) *Completing the Inventory of the Solar System*. ASP Conference Series, **107**.

* Robinson, K.S. (1991) *Red Mars*. Harper Collins. (1994) *Green Mars*. Harper Collins. (1996) *Blue Mars*. Harper Collins.

• Sagan, C. (1980) *Cosmos*. New York: Random House.

• Savage, M. (1994) *The Millenial Project*. Boston: Little, Brown and Company.

• Servier, J. (1967) *L'Histoire de l'utopie*. Paris: Gallimard.

+ Smith, S. & Zubrin, R. (eds.) (1996) *Islands in the Sky*. New York: John Wiley.

+ Zubrin, R. & Wagner, R. (1996) *The Case for Mars*. London: Simon and Schuster. An excellent presentation of current projects to conquer Mars.

• Wilford, J. (1990) *Mars Beckons: The Mysteries, the Challenges, the Expectations of our Next Great Adventure in Space*. New York: Knopf.

* Wuckel, D. (1981) *Science Fiction*. Editions Leipzig.

+ Special Issue (1991) Terraforming. *Journal of the British Interplanetary Society*, **44**, 146–191.

+ Special Issue (1993) Terraforming. *Journal of the British Interplanetary Society*, **46**, 291–322.

+ Special Issue (1994) Lunar industrialisation and colonisation. *Journal of the British Interplanetary Society*, **47**, 516–554.

+ Special Issue (1995) Lunar based astronomy. *Journal of the British Interplanetary Society*, **48**, 67–108.

+ Special Issue (1995) Mars exploration. *Journal of the British Interplanetary Society*, **48**, 287–324.

+ Special Issue (1995) Terraforming. *Journal of the British Interplanetary Society*, **48**, 407–443.

+ Special Issue (1997) Economics of space commercialisation. *Journal of the British Interplanetary Society*, **50**, 43–86.

CHAPTER 2

* Anderson, P. (1970) *Tau Zero*. Garden City, New York: Doubleday.

+ Andrews, D. (1996) Cost considerations for interstellar missions. *Journal of the British Interplanetary Society*, **49**, 123–128.

* Asimov, I. (1982) *Foundation. Foundation and Empire. The Second Foundation*. Ballantine Books.

• Beardsley, T. (1999) Ways to go in space. *Scientific American*, February issue, 61.

• Brener, R. (1982) *Contact with the Stars*. New York: W.H. Freeman.

+ Bussard, R. (1960) Galactic matter and interstellar flight. *Astronautica Acta*, 6, 179–194.

+ Celnikier, L. (1993) *Basics of Spaceflight*. Paris: Editions Frontières. An excellent textbook on the physics of interstellar flight.

* Clarke, A. C. (1968) *2001, A Space Odyssey*. (1982) *2010, Odyssey Two*.

(1987) *2061, Odyssey Three*. (1997) *3001, The Final Odyssey*. Ballantine Books. Clarke's best known work has become a science fiction classic.

* Clarke, A. C. (1976) *Rendezvous with Rama*. Ballantine Books.

* Clarke, A. C. (ed.) (1990) *Project Solar Sail*. Penguin Books. An interesting collection of technical essays and science fiction short stories on the subject of solar sail boats. In particular, it contains the original 1953 story *The Wind from the Sun* by Clarke himself (pp. 18–49).

+ Crawford, I. (1990) Interstellar travel: a review for astronomers. *Quarterly Review of the Royal Astronomical Society*, **31**, 377–400.

+ Dyson, F. (1968) Interstellar transport. *Physics Today*, October issue, 41–47.

• Dyson, F. (1988) *Infinite in all Directions*. Perennial Library. Reflections on our cosmic future by one of the century's most original thinkers.

• Finney, B. & Jones, E. (eds.) (1985) *Interstellar Migration and the Human Experience*. Berkeley: University of California Press. The best available collection of essays on various technical, philosophical and sociological aspects of mankind's expansion across the Universe.

* Forward, R. (1984) *The Flight of the Dragonfly*. Pocket Books.

• Forward, R. (1988) *Future Magic*. New York: Avon Books.

• Forward, R. & Davies, J. (1988) *Mirror Matter: Pioneering Antimatter Physics*. New York: Wiley Science Editions.

• Freedman, L. (1992) *Voiliers de l'Espace*. L'Etincelle.

+ Goldsmith, D. (ed.) (1980) *The Quest for Extraterrestrial Life*. Mill Valley, California: University Science Books.

* Harrison, H. (1979) *Captive Universe*. Le Masque.

+ Landis, G. (1997) Small laser-pushed interstellar probe. *Journal of the British Interplanetary Society*, **50**, 149–154.

+ Mallove, E. & Matloff, E. (1989) *The Starflight Handbook*. New York: John Wiley. A standard reference on technical aspects of interstellar travel, widely consulted in Chapter 2.

+ Matthews, R. (1994) The close approach of stars in the solar neighbourhood. *Quarterly Review of the Royal Astronomical Society*, **35**, 1–9.

* Pellegrino, G. & Zembrowski, G. (1996) *The Killing Star*. New York: Avon Books.

+ Project Daedalus Study Group (1978) Project Daedalus. *Journal of the British Interplanetary Society* supplement.

• Ribes, J. C. & Monet, G. (1990) *La vie extraterrestre*. Paris: Larousse.

• Schatzman, E. (1986) *Les enfants d'Uranie*. Paris: Le Seuil.

+ Sheldon, E. & Giles, V. (1983) Celestial views from non-relativistic and relativistic spacecraft. *Journal of the British Interplanetary Society*, **36**, 99–114.

+ Tipler, F. (1996) Travelling to the other side of the universe. *Journal of the British Interplanetary Society*, **49**, 313–318.

+ Zuckerman, B. & Hart, M. (eds.) (1995) *Extraterrestrials: Where are They?* 2nd edn. Cambridge: Cambridge University Press.

+ Special Issue (1984) World ships. *Journal of the British Interplanetary Society*, **37**, 243–304.

+ Special Issue (1986) Interstellar studies. *Journal of the British Interplanetary Society*, **39**, 379–409.

• Special Issue (1994) Life in the Universe. *Scientific American*.

• Special Issue (1996) Planètes extrasolaires. *La Recherche*, 42–60.

CHAPTER 3

• Berry, A. (1974) *The Next Ten Thousand Years*. Jonathan Cape Ltd.

+ Boothroyd, A., Sackmann, J. & Kraemer, K. (1993) Our Sun: present and future. *Astrophysical Journal*, **418**, 457–471.

• Carusi, A. (1995) Astéroïdes et comètes: les menaces sur la terre. *Pour la Science*, **212**, 90–97.

+ Chapman, C. & Morisson, D. (1994) Impact on the Earth by asteroids and comets. *Nature*, **367**, 33–40.

* Clarke, A. C. (1956) *The City and the Stars*. Harcourt, Brace and World.

* Clarke, A. C. (1993) *The Hammer of God*. Bantam Spectra Book.

• Close, F. (1989) *End*. London: Simon and Schuster.

+ Cohen, J. (1995) Population growth and Earth's human carrying capacity. *Science*, **269**, 341–346.

+ Dyson, F. (1960) Search for artificial stellar sources of infrared radiation. *Science*, **131**, 1667.

+ Gehrels, T. (ed.) (1995) *Hazards Due to Comets and Asteroids*. University of Arizona Press.

+ Gott Jr, R. (1993) Implications of the Copernican principle for our future prospects. *Nature*, **363**, 315–319.

* Hoyle, F. & Hoyle, G. (1973) *Inferno*. William Heinemann.

• Leslie, J. (1996) *The End of the World*. London: Routledge.

• Lewis, J. (1996) *Rain of Iron and Ice*. Helix Books.

* Niven, L. (1972) *Ringworld*. London: Victor Gollancz.

• Reeves, H. (1983) *Atoms of Silence*. MIT University Press; and (1994) Toronto: General Publishing.

+ Sleep, K., Zahnle, J., Kasting, O. J. & Morowitz, H. (1989) Annihilation of ecosystems by large asteroid impacts on the early Earth. *Nature*, **342**, 139–142.

*Stapledon, O. (1968) *Star Maker*. First published in 1937. *Last and First Men*. First published 1931. Dover Publications.

+Taube, M. (1981) Future of the terrestrial civilisation over a period of billions of Years. *Journal of the British Interplanetary Society*, **35**, 219–226.

+Taube, M. (1985) *Evolution of Matter and Energy on a Cosmic and Planetary Scale*. Springer Verlag. A concise discussion of the material resources of our planet and constraints on human expansion into the Solar System and the Galaxy.

*Van Herp, J. (1973) *Panorama de la science-fiction*. Belgium: Ed. Gerard and Co.

*Wells, H. G. (1995) *The Science Fiction*, vol. 1: *The Time Machine, The War of the Worlds, First Men in the Moon*. Phoenix.

•Special Issue (1996) L'Extinction des dinosaures. *La Recherche*, 293, 53–69.

CHAPTER 4

+Adams, F. & McLaughlin, G. (1997) A dying universe: the long term fate of astrophysical objects. *Reviews of Modern Physics*, **69**, 337–353. A revised and corrected version of the Dyson article cited below.

+Barrow, J. & Tipler, F. (1978) Eternity is unstable. *Nature*, **276**, 453–459.

+Barrow, J. & Tipler, F. (1986) *The Anthropic Cosmological Principle*. Oxford: Oxford University Press. A standard reference for any scientific discussion concerning mankind's place in the Universe.

+Basu, B. & Lynden-Bell, D. (1990) A survey of entropy in the universe. *Quarterly Review of the Royal Astronomical Society*, **31**, 359–369.

•Davies, P. (1994) *The Last Three Minutes*. USA: Basic Books.

+Dyson, F. (1979) Time without end: physics and biology in an open universe. *Reviews of Modern Physics*, **51**, 447–460. The founding article in the area of scientific eschatology, now a classic.

+Frautschi, F. (1982) Entropy in an expanding universe. *Science*, **217**, 593–599.

•Gribbin, J. (1987) *The Omega Point*. London: Heinemann.

*Hoyle, F. (1960) *The Black Cloud*. Penguin Books.

•Islam, J. (1983) *The Ultimate Fate of the Universe*. Cambridge: Cambridge University Press.

+Krauss, L. & Starkmann, G. (1999) Preprint (Astrophys/9902189).

•Luminet, J. P. (1987) *Black Holes*. Cambridge: Cambridge University Press.

•Montmerle, T. & Prantzos, N. (1988) *Soleils éclatés (les supernovae)*. Paris: Presses du CNRS–CEA.

*Poe, E. A. (1987) *Eureka*. In *The Science Fiction of Edgar Allan Poe*. Penguin Books.

•Prantzos, N. & Cassé, M. (1984) L'Avenir de l'univers. *La Recherche*, **15**, 839–847.

•Prantzos, N. (1987) Toi, l'univers, quand tu mourras... *Ciel et Espace*, **219**, 50–56.

•Rees, M. (1997) *Before the Beginning: Our Universe and Others*. London: Simon & Schuster.

•Reeves, H. (1991) *Hour of our Delight: Cosmic Evolution, Order and Complexity*. New York: W.H. Freeman.

•Reeves, H. (1996) *Last News from the Cosmos*. Toronto: General Publishing.

•Silk, J. (1997) *The Big Bang*, 2nd edn. New York: W. H. Freeman.

•Thuan, T. X. (1988) *La mélodie secrète*. Paris: Fayard.

+Tipler, F. (1994) *The Physics of Immortality*. London: Doubleday.

*Zembrowski, G. (1980) *Macrolife*. London: Futura Publications.

INDEX